Sam Mitchell
Fault-Tracing: Against Quine-Duhem

Epistemic Studies

Philosophy of Science, Cognition and Mind

Edited by
Michael Esfeld, Stephan Hartmann, Albert Newen

Editorial Advisory Board:
Katalin Balog, Claus Beisbart, Craig Callender, Tim Crane, Katja Crone,
Ophelia Deroy, Mauro Dorato, Alison Fernandes, Jens Harbecke,
Vera Hoffmann-Kolss, Max Kistler, Beate Krickel, Anna Marmodoro, Alyssa Ney,
Hans Rott, Wolfgang Spohn, Gottfried Vosgerau

Volume 40

Sam Mitchell
Fault-Tracing: Against Quine-Duhem

A Defense of the Objectivity of Scientific Justification

DE GRUYTER

ISBN 978-3-11-099678-4
e-ISBN (PDF) 978-3-11-068504-6
e-ISBN (EPUB) 978-3-11-068509-1
ISSN 2512-5168

Library of Congress Control Number: 2020944016

Bibliographic information published by the Deutsche Nationalbibliothek
The Deutsche Nationalbibliothek lists this publication in the Deutsche Nationalbibliografie;
detailed bibliographic data are available on the Internet at http://dnb.dnb.de.

© 2022 Walter de Gruyter GmbH, Berlin/Boston
This volume is text- and page-identical with the hardback published in 2020.
Typesetting: Integra Software Services Pvt. Ltd.
Printing and binding: CPI books GmbH, Leck

www.degruyter.com

Contents

Introduction — 1
1   The Quine-Duhem hypothesis — 1
2   Independent confirmation — 6
3   The key theses of Constructivism — 12
4   Outline of the course of this book — 18

1   An Example of a Constructive Tree in Darwin — 21
1.1   Isolated alpine populations have a recent common ancestor — 21
1.2   Explanation of the example — 26
1.2.1   The root node — 26
1.2.2   Confirming the auxiliaries — 27
1.2.3   The leaves — 28
1.3   Can this example be made complete? — 33
1.4   Why are examples of a constructive tree not more common? — 35
1.5   When should we expect to see scientists (and others) giving constructive trees, or parts of them? — 36

2   The Meaning of Independent Confirmation — 39
2.1   Observation independence and hypothesis independence — 40
2.2   Violations of observation independence: Examples — 41
2.3   Violations of hypothesis independence: Examples — 44
2.4   How to confirm auxiliaries independently using Bayesianism — 47
2.5   Constructive trees give the relevant background knowledge for a justification — 49
2.6   Cycles and why they are forbidden — 51
2.7   Examples of cycles in the literature — 53
2.8   An Example: Newton's original idea for absolute velocity — 55

3   How to Confirm Hypotheses about Unobservables — 59
3.1   Glymour's idea — 61
3.2   Adapting Glymour's idea — 63
3.3   A toy example — 64
3.4   Various objections and replies — 68
3.4.1   Constructivism is contrived — 68

3.4.2	The objection from van Fraassen's work —— 69	
3.4.3	Constructivism is too permissive —— 69	
3.4.4	The example takes for granted many hypotheses, so it isn't really a real confirmation at all —— 70	
3.4.5	These initial steps are a very poor justification —— 71	
3.4.6	Justification only begins holistically —— 71	
3.5	The scales of justice —— 72	
3.6	Apparatus —— 73	
3.7	The theory —— 74	
3.8	List of observables —— 75	
3.9	Confirming that two intersections are equidistant from the fulcrum —— 76	
3.10	Confirming two bodies have identical weight —— 77	
3.11	Establishing a scale: Integer multiples and fractions of weight —— 78	
3.12	Confirming a scale of length —— 80	
3.13	Overview of the balance —— 81	
4	**Fault Tracing —— 83**	
4.1	A nasty surprise —— 83	
4.2	What might have gone wrong? —— 83	
4.3	Independently confirming that that is what went wrong —— 85	
4.4	In some cases, the observations prevent a choice of blame —— 87	
	Appendix: What about Mercury? —— 89	
	The constructive answer to the example —— 90	
5	**Examples of Fault Tracing —— 93**	
5.1	First example: Polarized sunglasses —— 93	
5.1.1	First stage: Observations are inconsistent with beliefs —— 94	
5.1.2	Second stage: Make an intelligent guess at where the fault lies —— 95	
5.1.3	Third stage: Formulate a theory that avoids conflict —— 96	
5.1.4	Fourth stage: Confirming the diagnosis —— 97	
5.2	Second example: Compton's description of fault tracing in measuring X-ray wavelengths —— 97	
5.2.1	Background —— 97	
5.2.2	Bragg diffraction —— 98	
5.2.3	The problem, and initial attempts at a solution —— 99	
5.2.4	Trying to discover the source of the error —— 100	

5.2.5	Restoring agreement with the data —— 103	
5.3	Third example: Tracing errors to the experimenter —— 104	
5.3.1	Purposes of this section —— 104	
5.3.2	An interesting and simple example —— 104	
5.3.3	Tracing the fault —— 106	
5.4	Fault-tracing in natural science —— 107	
5.4.1	Synopsis of fault-tracing —— 107	
5.4.2	Closing words: The reply to reductionism and verificationism —— 108	

6 Chang's Paradox —— 111

6.1	Chang's paradox and its solution —— 111
6.1.1	The paradox —— 111
6.1.2	Initial steps in measuring temperature —— 113
6.1.3	Establish that mercury thermometers measure temperatures —— 114
6.2	Ordinal and linear scales —— 117
6.2.1	The course of experiments —— 118
6.2.2	What must be taken for granted? —— 120

7 Escaping Cycles of Confirmation —— 125

7.1	A cycle in Copernicanism —— 126
7.1.1	Copernicus had independent constructive evidence for each of the three hypotheses —— 129
7.1.2	An independent justification for heliocentrism —— 133
7.1.3	Amplifying justifications —— 133
7.2	Other ways that Copernicanism illustrates the superiority of constructivism —— 135
7.2.1	Other instances of constructive confirmation in Copernicus —— 137
7.2.2	It is very difficult to square Copernicus with different data —— 139
7.2.3	The advantage of the Copernican system —— 140

8 The Theory-Ladenness of Observation —— 145

8.1	The criticism from the theory-ladenness of observation —— 146
8.2	Theory-ladenness must allow for unexpected outcomes —— 147
8.3	We can distinguish between outcomes without justifying their necessary conditions —— 148

8.4	Why constructive trees justify without a justification for backing-hypotheses — 152	
8.5	Replies to objections — 154	
9	**Cycles of Observations — 161**	
9.1	Why constructivism gives a better account of observation — 161	
9.2	Examples of vicious background-dependence in observations — 163	
9.3	The literature on the topic — 165	
9.4	Chalmers' example; the constructivist analysis — 165	
9.5	Conclusion — 167	
10	**Van Fraassen's Paradox — 169**	
10.1	Van Fraassen's paradox — 170	
10.1.1	Why this is a paradox — 171	
10.2	The reply to the paradox — 172	
10.2.1	What is different about observation — 173	
10.2.2	The problem with van Fraassen's paradox — 176	
11	**Human Values Are Irrelevant to Empirical Justification — 179**	
11.1	Why does Quine-Duhem matter to philosophy? — 179	
11.2	Scientific change does not require pragmatic virtues — 180	
11.3	Justified empirical beliefs do not depend upon social values — 183	
11.4	Does constructivism really avoid pragmatic virtues? — 185	
12	**From Constructivism to Metaphysics: Potential Applications — 191**	
12.1	Empiricism — 191	
12.2	Possibility without possible worlds — 198	
12.3	A case for intuitionist logic and bottom-up metaphysics — 204

Bibliography — 211

Name Index — 217

Subject Index — 219

Introduction

1 The Quine-Duhem hypothesis

To anyone even slightly acquainted with the history of philosophy since 1950, the view defended in this book must seem utterly untenable. It argues against the position on justifications in the natural sciences that W. V. O. Quine, drawing upon the work of Pierre Duhem, so decisively championed in the mid-twentieth century. The upshot of the debate was a decisive victory for the *Quine-Duhem hypothesis* – the view that *no matter what outcomes we have observed at any time, the practice of science cannot use only those outcomes to select whether to give up or to adopt some given hypothesis*. We may believe in spite of an apparent refutation, by blaming some other hypothesis instead. By contrast, this book argues that the outcomes of observation sometimes do not permit us any choice in deciding which hypotheses are at fault when we observe something that our former beliefs prohibited.

An academic book should be an attempt to make a move in the way we now think about some subject. It must begin with the position in that subject as it is widely perceived at present before it attempts to move forward. This book, then, is aimed at a readership that is familiar with views on confirmation in the philosophy of science.

I do *not* want to go back to the battles of the 1950s and 1960s. The literature is unmanageably massive and framed in ways that are now dated. Revisiting it will not change anyone's mind. That literature is almost exclusively concerned with Quine's *first* dogma – the analytic/synthetic distinction. That focus continues even up to the present day (Chalmers 2011; Pickel & Schulz 2018).

This book, by contrast, looks closely at Quine's *second* dogma, namely that hypotheses of a scientific theory face observations not individually, but only as a collective group. This hypothesis has been much more widely seen as true and has had a huge influence (although Clark Glymour did politely demur to some extent (1980)). In the sense in which it is true, I will argue, the Quine-Duhem hypothesis does not follow from it. It has, moreover, had pernicious effects. It has yielded a poor analysis of justification as it occurs within natural science.

What is the argument for Quine-Duhem? The second dogma says that we may compensate for the future outcomes of observation, not by altering our view of the justification for that given hypothesis, but rather by adapting our view on the justification of other hypotheses we believe.

Quine then presented a dilemma:
1. Either empirical justifications depend upon auxiliaries, which must in turn be justified, to give a circle or endless sequence, or not.
2. If empirical justifications do ramify endlessly, or cycle, then the Quine-Duhem hypothesis is true.
3. If they don't, then some version of foundationalism is true.
4. Foundationalism is hopeless.
5. So the Quine-Duhem hypothesis is true.

This book shows that 3 is false. The chains of justification for a succession of auxiliaries can come to an end, in a finite time, without precipitating us into any version of foundationalism. Observation can be as theory-laden as you like. We can keep the examples of shifting blame to auxiliaries that we find in the history of science or our own reasoning. Empirical results are indeed widely interconnected; observations of X-ray diffraction do have consequences for hypotheses about the lift generated by the wing of a bird. The sense-data theory of perception can be utterly hopeless. We can be as anti-reductionist as we like. We can gain the advances for which Quine argued – the rejection of foundationalism and the sensitivity to auxiliary hypotheses – without surrendering to the pragmatist view of science that he so vividly presented in the last section of "Two Dogmas of Empiricism" (Quine 1953 (1980)). We can successfully practice and reason about the science we in fact possess without the Quine-Duhem hypothesis. We can be empiricists. That is the view that seems utterly untenable but which this book defends.

I call this alternative to the Quine-Duhem hypothesis *constructivism* because there is a clear intuitive sense in which the justification for a hypothesis is constructed from outcomes of observation. In nearly every case, a justification uses auxiliary hypotheses, and these must be confirmed independently of the target hypothesis. The process must terminate in hypotheses that are justified from the outcomes of observation without using additional auxiliaries. The word 'constructive' is intended to be analogous to a constructive proof in logic or mathematics. Constructivism in the philosophy of science has close connections to intuitionism in the philosophy of mathematics (Mitchell 2003). The clearest descriptive title for the view would be 'empiricist constructivism', but that is much too close to van Fraassen's constructive empiricism (1980, 1989). Unlike constructive empiricism, constructivism holds that hypotheses concerning unobservable entities are justified by the outcomes of observation in the same way as hypotheses concerning observable entities.

In "Two Dogmas of Empiricism," Quine (1953 (1980), 43) wrote:

> Any statement can be held true come what may, if we make drastic enough adjustments elsewhere in the system.

This is the Quine-Duhem hypothesis. The view occurs elsewhere. Two influential statements of it are:

> The physicist can never subject an isolated hypothesis to experimental test, but only a whole group of hypotheses; when the experiment is in disagreement with his predictions, what he learns is that at least one of the hypotheses . . . is unacceptable . . . but the experiment does not designate which one should be changed. (Duhem 1914 (1962), 187, see also 185)

> In principle, it would always be possible to retain [a hypothesis] even in the face of seriously adverse test results – providing we are willing to make sufficiently radical and perhaps burdensome revisions among our auxiliary hypotheses. (Hempel 1964, 28).

This viewpoint has been endlessly repeated up until the present day, like an incantation in a dead tongue. What follows is a sample, which I do not claim to be exhaustive: Braithwaite 1953, 19; Suppe 1977, 75; Hudson 1994, 606; Leplin 1997, 155; Hacking 1999, 71; Friedman 2001, 72; Longino 2002, 63; Collins 2004, 100; Worrall 2010, 127; Hudson 2014, 48. The point was even conceded as literally true by those who thought its consequences had been overdrawn (Popper 1963, 239; Grünbaum 1960). I have noticed increasing ambivalence and disquiet on the subject (see, for example, Kitcher 2001, 36), but we keep believing it, or saying that we do.

The Quine-Duhem hypothesis is important for the broader topic of epistemology, not just for the philosophy of science. Still, it is important that it should be answered from within the philosophy of natural science. That is where the argument was first proposed and argued. Duhem was a philosopher of science, and made his case from within that discipline. Quine was explicitly thinking in terms of natural science. Besides providing the arguments that first supported the Quine-Duhem hypothesis, natural science ought to play a particularly influential role in epistemology more generally. It provides a rich store of detailed and explicit examples of *a posteriori* justification, and it is hard to maintain that we do not know some of the hypotheses which these justifications recommend.

In addition, some of the most appealing examples of the Quine-Duhem hypothesis come from natural science. Thomas Kuhn gave a fund of examples (1962 [1996]). Newton could not, initially, account for the orbit of the moon (30, 81). Mercury didn't fit Newton's laws (81). The Copernican hypothesis was at odds with our inability to detect parallax in the fixed stars (26). Just as the Quine-Duhem hypothesis says, we did not blame Newton for all this, but some auxiliary instead. In every example but Mercury, the result was an eventual triumph for Newtonian mechanics. Examples such as these are commonplace in all areas of science. The Quine-Duhem hypothesis gives a nice account of our refusal to surrender the main hypothesis in the light of these difficulties. Scientists simply blame some auxiliary instead of their favored hypothesis.

While these examples have a great deal of intuitive appeal, we should remember that the Quine-Duhem hypothesis is a universal generalization. It alleges that it *always* possible to find more than one diagnosis for some set of data that challenges the hypotheses which the scientific community endorses. That is simply not true.

The Rayleigh-Jeans (or 'ultraviolet') catastrophe must be avoided by abandoning the independence of frequency and energy. Rayleigh and Jeans derived an expression for the radiation from a cavity based rigorously on classical physics. It dramatically failed to fit the data from high frequencies. Planck derived a different law, but needed to make the energy of the entities contributing to the radiation proportional to their frequencies. That was a deep and dramatic break from classical physics, but it has proved to be unavoidable. Given the observations we now possess, nobody has yet proposed an alternative that survives the evidence.

Another example concerns the Michelson-Morley experiment. We can detect differences in the speed of light along different directions by splitting a beam and sending light down two paths of equal length at right angles, and then reflecting it back again. Rotating the apparatus should result in an observable shift in the interference fringes if light travels faster in one direction than another. There is no such shift. There were many theories in 1900 that yielded this null result, but the only ones that now remain as live options include both time dilation and length contraction. No other way succeeds, given the totality of our observations.

A third example: Lord Kelvin objected to Darwin by arguing that the earth's internal temperature shows that it is younger than Darwin required. A world that young would not have allowed enough time for all the species to evolve, given how slowly they reproduce. We now know that Kelvin assumed, falsely, that atoms in the earth released no energy over time (as they do by radioactive decay). So the earth is older than its internal temperature would indicate without radioactivity. No other solution survives, and we arrived at this one by empirical investigation.

Our current evidence strongly supports the hypothesis that frequency and energy are not independent, that both time dilation and length contraction are required for the round-trip experiments with light, and that some atoms in the earth release energy over time as they decay radioactively. Constructivism insists that we have excellent *evidence* that the fault lay in *these* hypotheses, and did not lie in certain other hypotheses. The evidence that Einstein was right in accounting for the results by including both length contraction and time dilation in special relativity is overwhelming, and every attempt to deny this has clashed with some outcome we have observed.

There must be, within science as it is practiced, some way to use the outcomes of observation to discover whether or not some hypothesis is at fault when the evidence challenges the hypotheses we formerly believed. Constructivism proposes a mechanism for this, *fault tracing*.

To anticipate briefly: We have self-contained justifications, constructed from the evidence, which can all be brought into agreement with each other by convicting some hypotheses and acquitting others. We do the best we can, given the evidence, to get these independent justifications to agree. Sometimes – as in the above examples – the process is highly successful, and we know which belief was wrong, why it was wrong, and how well-justified hypotheses correct for it. Given those observations, we cannot hold on to any hypothesis we like; the evidence points decisively towards some hypotheses and away from others. At other points in history, or given some sets of evidence, we cannot yet locate the fault. In those cases, we can do what the Quine-Duhem hypothesis says we can always do; we can reasonably locate the blame at different places.

The only argument for Quine-Duhem is the sketch given in the above quotations. It claims that confirmation is always relative to auxiliaries. We can always blame those auxiliaries when the evidence supports something we want to avoid. So the Quine-Duhem hypothesis follows. Both Richard Creath and Clark Glymour noted that there is a gap in Quine's work when it comes to giving a detailed account of the confirmation theory in "Two Dogmas" (Glymour 1980, 6; Creath 1990, 20). "Two Dogmas" itself cites Duhem (1953, 41). There is an argument first given in Hempel's work (1966) that fills out many of the claims Quine makes, and which parallels Duhem. (Carnap was first to see in Hempel a clear and explicit version of Quine's argument (Carnap 1966, 269).)

The theory of confirmation on which Hempel's argument depended has been bypassed by subsequent developments. Hempel viewed confirmation as simply the logical entailment of an observation from the target hypothesis, together with auxiliary hypotheses (1966, 23–25). Then, when the negation of the prediction is observed, *modus tollens* allows us to say that at least one of the hypotheses and the auxiliaries must be false, but it doesn't allow us to single out which. Pierre Duhem gives the same argument, but focuses more on the theory-ladenness of the observations involved (Duhem 1914 (1962), 183–190; Laudan 1965). This argument depends upon an unaugmented hypothetico-deductive view of confirmation. That is by no means the only view of confirmation available at the moment, and it has some particularly troublesome problems (Glymour 1980, 29–48; Glymour 1980hd). The upshot is that aside from the examples from the history of science, what little argument there was for the Quine-Duhem hypothesis depended upon an early theory of confirmation that nobody can now embrace without qualification.

This book defends the idea that we can avoid the Quine-Duhem hypothesis by appealing to fault-tracing. In doing so, though, I hope to make a stronger case plausible. There is a good case that constructivism gives a *better* account of empirical justification, not just an alternative one. The pragmatic picture that Quine gave claimed that fault-tracing was impossible, though there are apparent examples of it. Constructivism gives an account of fault-tracing, so that we do not have to somehow deny the examples. It has at least one additional advantage too. It gives a way for empirical investigation to provide *independent* confirmation for hypotheses. There are considerable difficulties for Quine's pragmatic picture under this heading.

2 Independent confirmation

In addition to examples of fault-tracing, the Quine-Duhem hypothesis makes it impossible to understand how we could confirm hypotheses by two independent methods. This is something that we clearly do.

Aaron Edidin divided the problems that a theory of confirmation ought to solve into two:
(1) When does a piece of evidence confirm a hypothesis relative to a set of auxiliary assumptions?

And
(2) When does such relative confirmation genuinely contribute to the credibility of the hypothesis? (Edidin 1988, 266)

I will call the first question *relative confirmation*, and the second *real confirmation* (following Edidin).

One problem of confirmation, then, is this: what are the conditions under which *relative* confirmation, confirmation of H by $\{E\}$[1] relative to a set of background auxiliaries $\{A\}$, turns into *real* confirmation, when observing that $\{E\}$ ought genuinely to justify believing that H?

---

1 Throughout this book I adopt the convention of representing a set of, for example, auxiliary hypotheses as $\{A\}$, and an arbitrary element as A. The notation is extended in the obvious way, so that a set of observations is $\{E\}$, and an arbitrary element is E, etc.

David Christensen considered the matter carefully (Christensen 1997). He observed first that we cannot take the observation to confirm a hypothesis relative to *some background auxiliaries or other*. For, if we did that, then any observation will confirm any hypothesis, since according to each theory of confirmation, there are always *some* auxiliaries that confirm that hypothesis relative to any observation. (Darwin's finches confirm that Mars has an elliptical orbit relative to the auxiliary that, if a tool-using bird exists, then the orbit of Mars is roughly elliptical.)

We could try saying that only *true* auxiliaries genuinely confirm their target hypothesis. But theories of confirmation have an inevitably epistemological component. Unless there is some theory of how we know which auxiliaries are true, we cannot use the theory of confirmation to address the issue of how *we* are justified by the evidence in believing something. We cannot depend upon true auxiliaries unless we already know how the evidence shows that a hypothesis is true, which is the problem that real confirmation is supposed to address. (And besides, as Grimes (1987) noticed, it is in fact true that if a tool-using finch exists, then Mars has an approximately elliptical orbit, because Mars does.)

Nor can we just say that confirmation is genuine if it is relative to auxiliaries that are *genuinely believed* by the individual (or his community), for the obvious reason that sincerely held but completely unjustified beliefs will support utterly bizarre instances of confirmation, which will then count as genuine. (Think here of the observations that "confirm" that the apocalypse is at hand, or that aliens once visited us, or any number of other fanciful views. The sincerity of their adherents is not in doubt.) Confirmation is a matter of what one *ought* to believe in the light of the evidence, and we are well aware that, even in our own case, we sometimes make mistakes about it. We do in fact discriminate between instances of confirmation we take to be reasonable, and those we do not. We do not do so by looking at the sincerity of those who espouse the justifications, nor even, sometimes, at whether we share their beliefs.

The most plausible suggestion, then, is that relative confirmation is genuine only if the background hypotheses are themselves justified. Given the demise of *a priori* theories of the justification of empirical hypotheses, Christensen concludes, as I do, that there must be *evidence* justifying the auxiliaries if any instance of relative confirmation is to count as acceptable.

But now, clearly, the same problem arises again for the auxiliaries $\{A\}$ as originally arose for the target hypothesis H. Not all cases of $\{A\}$ will actually work in justifying H when E is observed. What distinguishes those that do from those that don't? Well, those that do are cases where each A is confirmed. But A must be confirmed relative to new auxiliaries $\{A^*\}$, and not all instances of $\{A^*\}$

will work. What distinguishes those that do from those that don't? We are obviously back at exactly the question we began with.

In real confirmation, when $\{E\}$ confirms H relative to $\{A\}$, a full account of the fact that one ought to believe H must include reasons justifying the use of the elements of $\{A\}$. Unless every A has to be confirmed, the confirmation theory will be subject to counterexamples in which absurd hypotheses are confirmed by true observations relative to absurd auxiliaries. But if $\{A\}$ does have to be confirmed, then the justification for $\{A\}$ is a necessary condition for the justification that $\{E\}$ provides for H. It cannot be omitted from the justification of H without undermining it, since if any A were *un*justified, H would be too.

What distinguishes real confirmation from relative confirmation is just that in real confirmation, $\{A\}$ must have a genuine justification from the evidence, and in relative confirmation it need not. In relative confirmation it is harmless that H is confirmed only if $\{A\}$ is secure, for that is just what relative confirmation explains. But plenty of hypotheses that are not justified would be justified if something else were true. We can be justified in believing them only if we somehow come to have a justified belief that the preconditions for some justification actually hold.

Horwich (1983), Edidin (1988) and Mitchell (1995) all make the following very plausible suggestion at this point in the argument:

> Observing evidence $\{E\}$ as opposed to refuting evidence $\{E'\}$ genuinely confirms H relative to $\{A\}$ if and only if every member of $\{A\}$ is *confirmed independently* of whether H is true or false, and independently of whether $\{E\}$ or $\{E'\}$ was actually observed (see Mitchell 1995, 244).

The trouble is, as Christensen noted, that apart from an outline in Mitchell (1995), nobody has defended any detailed theory of how to confirm one hypothesis independently of some other hypothesis, or how to give two independent justifications for a single hypothesis (Christensen 1997, 381). Constructivism does so.

Why think that independent justification should be possible at all? After all, if the Quine-Duhem hypothesis is true, then the "independent" confirmation of some auxiliary A will inevitably require another auxiliary A', and we will begin an infinite regress or circular justifications. Moreover, the ramifications of justification will spread without limit. If we attempt to give every empirical justification for A, we will depend upon a very large number of successor auxiliaries. If we then go on to attempt to give every justification for each of *them*, the same proliferation will recur. Eventually, then, "The unit of empirical significance is the whole of science" (Quine 1953, 42), and independent justification is impossible.

The obvious reason for thinking that independent justification must be possible is that there are a great many hypotheses for which we possess overwhelming real confirmation. That is, there are excellent reasons for believing them, and educated people in the sciences, and many scientifically literate people who are not practicing scientists, can tell you the observations that confirm them. Since one runs into conflict with different recorded observations as one tries to think of different criticisms of these hypotheses, there exists the very strong suggestion that they are massively independently confirmed. Human beings share a common ancestor with chimpanzees. The sun is closer to the center of mass of the solar system than the earth. Atoms exist. None of these were welcome pieces of news. All were stoutly resisted. All resistance gave way under the weight of evidence. That is the obvious thing to say, and it is to the advantage of constructivism that it can say it. If the Quine-Duhem hypothesis were true, we could hold onto their denials today without having to repudiate the things we have actually observed. Christensen's line of argument supports the view that if real confirmation is to be possible, then there is independent justification for auxiliaries. Real confirmation is possible for some hypotheses. So there is independent justification.

Nor is it difficult to give examples of independent justifications. When Adams and Leverrier predicted the orbit of a planet from irregularities in the orbit of Uranus, that hypothesis was independently confirmed by the sight of the planet using a telescope. Biogeography in the light of continental drift provides many instances of justification for hypotheses in evolution independently of the justifications that Darwin offered. The number of electrons needed to electrolyze one mole of silver gives an estimate of Avogadro's number. That estimate is independently confirmed by counting radiation from a radioactive element and correlating it to the mass of that element which decays over time.

All of the evidence for *every* auxiliary in these experiments, repeated successively for their auxiliaries, might (for all I know) eventually cover the whole of science, and prevent these justifications from being independent. But it doesn't follow that we cannot give *some* evidence for each auxiliary that is independent of the target hypothesis and the evidence. Then we would possess one item of real justification. The examples above each give more than one justification for a single hypothesis, which can be completed in a way that makes them independent. We will see that auxiliaries, too, can be reinforced by adding in justifications that are independent of the hypothesis they eventually support, so that a doubt about some auxiliary can often be answered. In some cases, real justification is overwhelming, and it is just false that we can regard the denial as justified in the light of the evidence.

The words 'real justification' signal only that the justification is not relative. A single real justification need not, by itself, be particularly convincing. (I will say that real justification is *overwhelming* when it becomes impossible to see how to maintain the denial of a hypothesis given the evidence.) We usually require many independent real justifications to reply to objections and make a hypothesis difficult to avoid. But a single piece of real confirmation is still a justification. The outcomes we actually observed came out that way, and not another. Because they came out that way, they justify the target hypothesis as opposed to refuting it. So it's a complete justification even if it falls short of conferring much conviction. It is begging the question to argue that, because everything connects to everything else, giving just some evidence is not really giving a justification.

Constructivism, then, begins with an instance of relative confirmation, H_O is confirmed by observing $\{E_O\}$, rather than some $\neg E_O$, using auxiliaries $\{A_O\}$. Then each A_O must be confirmed by new, independent, evidence, $\{E_1\}$, relative to new auxiliaries $\{A_1\}$. This process is brought to an end within a finite time, when some hypothesis A_n gets confirmed without any further auxiliaries. This is a *constructive tree*, and is one complete, finite, reason to believe H_O.

(This isn't intended as a temporal process of course. Constructivism doesn't allege that working scientists *first* begin with a hypothesis that is directly confirmed by the evidence, *and then* figure out how to independently confirm some other hypothesis using it, and so on. Constructive trees are predicted to be something we can devise when we see that a hypothesis is empirically justified. We ought to be able to reconstruct them when, for example, we contemplate why the scientific community is convinced of some hypothesis. We should be able to see how such a tree provides part of the background knowledge that some experiment or argument uses, and how accumulating such trees can fill in the reasons why some hypothesis is secure.)

We get strong independent confirmation when we can offer many of these constructive trees, using different hypotheses and sets of evidence, in support of a single target. For some hypotheses, we have such a large number of independent justifications that we simply cannot see how to defend $\neg H$ in the light of the evidence we possess. H is then overwhelmingly justified with respect to the observations that are involved in these independent justifications. Put intuitively, we cannot see "how it could be false", because every way we can devise for its falsehood runs into trouble with the evidence. The target, H, cannot appear in any of these constructive trees, for if it did so, we would not have confirmed some auxiliary in one of the justifications independently of the hypothesis which that justification was supposed to support.

That is the way that constructivism answers Quine's dilemma. When the outcomes of observation are sufficiently cooperative, we get overwhelming real confirmation that some hypotheses in an apparent conflict with the evidence are innocent, and overwhelming real justifications of the denials of others. So we cannot hang onto just any hypothesis regardless of what we observe.

2 Independent confirmation — 11

This idea isn't going to work unless we say something about what happens when we switch from one target hypothesis to another. For it is no use having a huge number of independent justifications for H if each and every one of these is easy to challenge. Suppose, in each of them, we use a unique hypothesis W (for 'weak'). Only very slight evidence can be mustered for W. When we make W into the target, and challenge it, it turns out to be very easily dispensable. Then, in spite of the many justifications for H, it could be justified only in a very speculative manner.

In natural science, we try to get a common set of hypotheses with few or no refutations, and independent confirmations that all agree within that common set. Each of our beliefs ought to be reasonably helpful in serving as auxiliaries to as many others as we can devise. It should not be easy to dismiss, as W was. This means we ought to be able at least to think of a way to confirm each belief independently of each particular hypothesis we use it to justify. If some hypothesis has no such justification, constructivism predicts that we ought to seek it assiduously. Constructivism has to say this, but we will see that fault-tracing, as a description of our actual practice, produces this result anyway. The puzzle-pieces fit together without forcing.

So a hypothesis is overwhelmingly justified when all (or almost all) constructive trees confirm it. But it turns out that constructive trees are very cheap to produce, and some very silly hypotheses can be confirmed by a few of them. Constructivism says that we rule out these silly hypotheses by demanding that we find a single set of highly justified hypotheses that keep each other's company in a mutually supportive way with respect to the outcomes of observation. But that raises the specter of the Quine-Duhem hypothesis again. If there is more than one set of such hypotheses, perhaps any hypothesis and its negation are both members of some collectively supportive gang.

Well, *perhaps* so. But constructivism shows how empirical justification can avoid the view that all justification is always relative. Since it denies this, there is no argument to the conclusion that it is *inevitable* that we can always keep an arbitrary hypothesis away from refutation or confirmation by the outcomes of experience. The argument from Duhem, that we could blame an auxiliary in the target justification, and could continue this policy without end, is gone. Since we do not have a general argument for underdetermination, we must look at the examples in the science we have before us. Can we find examples that make the Quine-Duhem hypothesis unbelievable? Can we find examples of mutually supportive hypotheses with enormous independent support from the outcomes of observation, so that one of them at least becomes overwhelmingly justified? Of course we can. Just try defending the idea that the center of mass of the solar system is closer to the earth than the sun.

This view is not foundationalist about the outcomes of observation. The outcomes of observation are not, as foundationalism alleges, atheoretical presentations of direct experience. Rather, they are judgments that our sensory apparatus distinguished an outcome that favored one theory-involved result over a different one. Observation can be as theory-laden as you like; we are still able to detect this difference. There is no good argument in the literature supporting the view that the demise of foundationalism prevents us from encountering refutations from outcomes of observation, and examples are all too obvious.

There are various different theories of relative confirmation in the literature – Bayesianism, Neyman-Pearson statistics and Likelihood theories are examples. Constructivism is not a rival to these contemporary theories of scientific justification. It shows how we may use any of them to move from relative to real justification. Any of these three versions of confirmation theory, I believe, can be formulated constructively. Constructivism requires only that a theory of relative confirmation be able to do two things. It must be able to confirm generalizations from observations of their instances, and it must be widely applicable to real examples of empirical reasoning. Any of the three can do these things.

All contemporary relative confirmation theories are sensitive to Clark Glymour's[2] *relevance* problem: evidence bears on some hypotheses more than others, and does not indiscriminately justify the entirety of our views simultaneously (Glymour 1980; Dorling 1979 (Bayesianism); Howson and Urbach 1989, 96 (Bayesianism); Mayo 1996, 456–458 (Neyman-Pearson statistics); Royall 1997 (Likelihood)). One might argue that, therefore, these relative theories already contain the ability to convict some hypotheses and acquit others when the evidence clashes with our beliefs. This is not so. In each case, the argument is that evidence can bear more on one hypothesis than another *given that certain auxiliaries are true*. Varying these given hypotheses varies the relevance relations. This is certainly an advance on hypothetico-deductivism, but it does not address the issue of where this given knowledge comes from.

3 The key theses of Constructivism

Here, I simply state the main theses of constructivism, so that the position is out in the open; later chapters offer deeper explanations and further arguments for them.

---

[2] I owe a great deal to Clark Glymour's work, particularly his (1980).

1. Real confirmation, not just relative confirmation, is needed for an adequate account of justification by observations in science. We know that the justifications we possess in natural science really do justify some hypotheses. We ought to be able to make this knowledge explicit and open to inspection.

2. Real confirmation can be given by showing that the auxiliary hypotheses of a proposed justification possess some justification that works independently of $\{E\}$ and H.

We have two justifications here. First, a target hypothesis, H, is justified by observing $\{E\}$ relative to an auxiliary hypothesis A. Call this the *target* justification of the two. In the second, *auxiliary*, justification, A is justified. The auxiliary justification must be independent of H and $\{E\}$ in the target justification.

So the auxiliary justification is supposed to answer the question "What reason is there to rely upon the truth of A when we use it to justify H upon observing $\{E\}$?" Such a reason must not ultimately depend upon H or $\{E\}$ being true (or probable, in some theories of relative confirmation), or it could not answer this question.

We contemplate two possibilities, observing $\{E\}$ or observing some $\neg E$. The first indicates that H is true, and the second does not. Why? Because there is a reason to believe other hypotheses, $\{A\}$ whether or not H or $\{E\}$ are true. Relative confirmation shows that if all $\{A\}$ are true, and $\{E\}$, this indicates that H is true. Under Constructivism, we have evidence that all the $\{A\}$ are true, because they've been justified in a way that we know will work whether or not H and $\{E\}$ are true (or probable). We know $\{E\}$ is true because we observed it. So we have evidence that H is true. That is how we get from relative to real confirmation.

3. The call for independent justification for A is a call for a justification that does not require, or assume, that H or $\{E\}$ possesses what I will call the *justifying virtue* of the theory of relative confirmation we are using.

Constructivism is a view about how we get from relative confirmation to real confirmation. Bayesianism, Neyman-Pearson statistics and Likelihood are all, as they stand now, varieties of relative confirmation. They differ about what happens when a target hypothesis is confirmed, and so differ in their justifying virtues. For Bayesianism, the justifying virtue is the probability of the target hypothesis. This is what increases for instances of confirmation, and decreases with refutation. So, for a Bayesian, we confirm auxiliary A independently of the target justification that uses it when we cite evidence increasing the probability of A without reference to the probabilities of H or $\{E\}$.

By number of publications, Bayesianism is the most popular contemporary theory of confirmation. Under that familiar theory, H is confirmed when and only when we observe $\{E\}$ such that the probability of H given that evidence is higher than the probability of H given contrary evidence. We should then update our subjective probability for H:

$$\Pr_{new}(H|\{B\}) = \Pr_{old}(H|\{B\})\, \Pr_{old}(\{E\}|H \wedge \{B\})/\Pr_{old}(\{E\}|\{B\})$$

The left-hand side is equal to $Pr_{old}(H|\{B\} \wedge \{E\})$. Making that substitution, the result is one statement of Bayes' theorem (Strevens 2012). What constructivism does is to provide an account of when we may move from $Pr_{new}(H|\{B\})$ to just $Pr_{new}(H)$; we may do this when we have some strong independent assurance of the justification for $\{B\}$.

I have picked Bayesianism as the implementation of a constructive theory of confirmation only because I must pick one relative theory of confirmation, Bayesianism looks to be the most popular, and it is computationally well-understood. I have already mentioned the fact that there are many other theories of relative confirmation: Deborah Mayo's (1996) error statistical theory, Clark Glymour's (1980) bootstrap theory, Likelihood theories (Royall 1997; Edwards 1972), other versions of Bayesianism, such as that proposed by Richard Jeffrey (1965), and defensible hypothetico-deductive theories (if there are any). I believe these would also serve as partners for constructivism, because constructivism makes no demands that are not already met by the existing theories of relative justification. We will see that constructivism requires confirmation of generalizations by their instances, and the ability to capture a range of intuitive examples from the history of science, science education, and occasionally philosophical invention. But all theories of relative confirmation can do these things.

4. From the above, we should be able to produce an independent justification for each auxiliary that we use in any proposed real justification. If those parent justifications depend upon new auxiliaries, then those new auxiliaries must in turn be justified by a grandparent justification that is independent of its descendants. This chain of justifications must come to an end.

For suppose it does not come to an end. It cannot then serve the purpose for which it is intended. That purpose was to provide us with evidence for each element of $\{A\}$, so that we know we can rely upon it to justify H from observing the outcomes $\{E\}$. One way of putting it is that we want to know that observing $\{E\}$ supports H *itself*, as opposed to being irrelevant to H, or providing some reason to doubt some A instead. If we cannot survey the chain of justifications, then we cannot possess this assurance. The only reasons we have for believing some A might, unknown to us, depend upon H already being true, or depend upon some E already being observed. (It might also depend upon some piece of background knowledge for which we possess no justification; see 6 and 7 below.)

5. So we must be able to produce, when we investigate intuitively compelling examples of empirical justification, a finite tree such that:
    a) each node is an instance of relative confirmation, except for the leaf nodes, which appeal to no auxiliaries and
    b) the root node is a proffered or target justification. The rest of the tree concerns a justification for the background knowledge upon which this justification depends.
    c) The edges represent the relation of: The parent node *provides a justification for an auxiliary that is used in* the child node. This justification in the parent node should work whether or not H, and should work whether or not any E is true.

This is a *Constructive tree*. I will shortly present some simple examples.

6. This structure prevents *cycles* of confirmation, where some hypothesis H is confirmed using an auxiliary A, and then A is confirmed *only* by appeal to H. The cycles can be extended so that H is confirmed using $A1$ which is confirmed using $A2$, which is confirmed using H.

All such cycles are prohibited as instances of real confirmation. This is because every constructive tree must give us a reason to believe an auxiliary whether or not the target hypothesis is true or believed. A cycle of justification cannot do so, as it ultimately depends upon the target hypothesis. So a cycle of justification is always legitimately rejected as a piece of real confirmation. That doesn't mean it cannot be suggestive, or have other uses, but it must be regarded as flawed as a justification by itself.

That is a surprising thing for constructivism to claim. Under the influence of Quine-Duhem, we are used to the idea that cycles of confirmation do give us a reason to believe. This book argues, though, that pure examples of cycles – where the example cannot be reformulated constructively – are both rare in natural science and strikingly unconvincing as reasons to believe an empirical hypothesis. Where a cycle looks superficially convincing, we know we can replace it with a constructive tree, escaping the cycle. Where we do not know this, the cycle is always legitimately rejected, and is strikingly odd. (So this surprising claim about empirical justification gets confirmed when we look at examples of empirical justifications.)

7. The branching chains of justification must terminate, after a finite number of steps, in justifications that depend only upon the fact that observations have had a particular outcome, without introducing additional auxiliaries in need of justification. These are the justifications at the leaves of the tree. No tree can "dead end" by depending upon some ultimate A that possesses no further justification. For if A is false, we wouldn't be justified in holding on to the hypotheses it is supposed to support. Leaves are all cases of justification by induction from observed instances.

8. A constructive tree provides one exhaustive justification for the ramifying chains of hypotheses we use as background for some given justification, the target justification. The target justification is simply the justification that we are interested in in this context, so that the same example can be a target in one tree, and the justification for an auxiliary in another. We are filling in one piece of background knowledge for this target by showing how the auxiliaries in it can be independently justified. When we find an example of a proffered justification in the practice of science, we ought to be able to show explicitly at least one way to construct a tree if constructivism is correct.

In examples familiar from science textbooks and papers, this target justification is all that is explicitly stated. Scientists share a common stock of background knowledge, so it would be pointless to go through it without some special reason. In some cases, though, we see more of the constructive tree. In the next chapter, we will see an example of Darwin going through the background knowledge for a justification in order to respond to an objection. Later on we will see how to use constructive trees to track down which auxiliary is in error.

9. We typically use several independent justifications to secure a single hypothesis in the light of the fallibility of any one of them.

Thus far, we have mostly been looking at ways to justify one hypothesis, A, by one justification that is independent of some separate justification for a second hypothesis, H. But it is also common in science to take a single hypothesis, H, and constructively justify it by two independent experiments. Each way of justifying it can be represented as a constructive tree. Some constructive trees justify a hypothesis independently of any hypothesis or observation in some other constructive tree. More often, there will be some sharing of the hypotheses used in the two justifications.

Each constructive tree traces a target hypothesis to a set of observed outcomes. Empirical justification is never incorrigible, so no constructive tree can confer certainty about the target hypothesis. This is something all theories of relative confirmation recognize and accommodate so that, for example in Bayesianism, no evidence can produce probabilities of one or zero for a hypothesis if the prior probability is not there already. But if there are a lot of constructive trees, and they all agree in justifying the same hypothesis, then it becomes harder and harder to see how the hypothesis could be false, since many independent things must have gone wrong, and in each case we have at least some evidence that it did not go wrong. As a result, no single constructive tree need be particularly convincing by itself, and yet a collection of them can provide good evidence for a single hypothesis which they each confirm.

Constructivism is neutral about whether some of these constructive trees, or some parts of them, play some special role in what expressions mean. It holds that each is fallible, and that each is refuted by evidence that differs from that which we actually observed. None of them, then, play the role that analytic hypotheses played in logical positivism. It might still be the case that some play a special role in the philosophy of language, for example in confirming the meaning or reference of some expression. But there is no need to speculate here. The aim of this book is restricted to the philosophy of science, and that is quite ambitious enough.

10. We seek two things in the hypotheses that are really justified. We seek a set that maximizes agreement among these different ways of confirming these hypotheses from the evidence. And we seek a set that minimizes conflict with the evidence. This book argues that these two objectives are not distinct. Tracing faults is one of the most important ways in which we maximize agreement between independent justifications. At least sometimes, seeking an informative theory and minimizing conflict with the evidence are not distinct activities.

By accumulation of constructive trees, we often get a cluster of hypotheses such that any one of the cluster is overwhelmingly confirmed. It is much better justified than its denial, so that we cannot see how to hold the denial true without either, first, preventing any evidence from justifying anything, or second, conflict with the outcomes of observation.

11. Constructive trees provide a sufficient condition for moving from relative justification to real justification. They are all we need to practice science and understand why natural scientists achieve consensus about what ought to be believed.

At various points in the book, I argue that constructivism might also be a necessary condition for real justification. The prohibitions constructivism holds to apply to scientific reasoning – particularly the ban on cycles of confirmation – are observed in the practice of natural science. But I do not have a positive argument to offer that constructivism must be the *only* way to act and reason scientifically. Perhaps there is some additional way to move from relative to real confirmation. So far, nobody has presented some other way to give a real justification though.

So the main thesis of this book is: *Constructivism is a sufficient condition for real confirmation, and could, so far as we know, also be a necessary condition for real confirmation.*

4 Outline of the course of this book

Constructivism can be viewed from a top-down or bottom-up perspective. The top-down perspective begins with a hypothesis, and then looks for constructive trees that justify it. The bottom-up perspective begins with outcomes of observation, and looks at how they confirm other hypotheses. In the former case, the worry is that we cannot get to the leaves of the tree, because we will always face justifications that depend upon auxiliaries. In the latter case, the worry is that, beginning with outcomes of observation, we will only be able to confirm hypotheses about observable entities.

Chapter 1 establishes that real scientists sometimes give sketches of constructive trees, or at least parts of them. It gives some examples of top-down initial segments of constructive trees. It leaves open, though, whether these trees can be given exhaustively, addressing every auxiliary, and all the way to the leaf nodes. Those issues get addressed in chapter 3, which gives an example of a bottom-up justification.

In chapter 2 I give a statement of constructivism in terms of a Bayesian theory of relative confirmation. The chapter says what it is for an auxiliary to be confirmed independently of a target hypothesis and the observations justifying it in the target justification. There are also examples of justifications containing cycles of confirmation, where it looks inevitable that auxiliaries cannot be confirmed independently. These are highly counterintuitive. They contrast to otherwise similar examples where we are aware of ways to prevent this cycling of justifications.

Chapter 3 looks at the way in which hypotheses concerning unobservable entities get constructively justified. I give two bottom-up examples of the way in which outcomes of observation result in justifying hypotheses about unobservables. Both examples concern length. They are necessarily primitive in appearance. The question being addressed is how empirical justification could get started, not how it performs once a large amount of background knowledge has already been confirmed. We see how constructivism makes the case that rulers and balances reveal facts about length and weight that we cannot observe without their aid.

Thus, there is a constructive justification for hypotheses about length from two independent sources, rulers and balances. We find that they do not agree about length in some cases. Chapter 4 shows how to use fault-tracing to resolve the disagreement and to locate the hypothesis that needs to be altered. We can confirm that this is the guilty hypothesis by getting independent evidence that it is in fact false.

That is a toy example. In chapter 5 I look at three more realistic examples of the same kind of fault tracing. The idea here is to show how widely applicable the strategy is. Chapter 6 extends the treatment of real examples by replying to an objection to constructivism from Hasok Chang (2004). Chang alleges that cycles of justification or fundamental principles accepted without evidence are inevitable and identifies some in the history of thermometry. The chapter shows, though, that it is possible to justify these hypotheses constructively, by making use of at least some of the features that Chang observes in the history.

Chapter 7 compares the kind of justifications that Copernicans could give for their view prior to the telescope to justifications for the Ptolemaic theory. Clark Glymour argued that there was a cycle in the Copernican justifications, but it is easy to see, both now and at the time, how to replace it with constructive justifications. It's much harder to see how to break out of the cycles that Ptolemy requires. I suggest, then, that the greater 'harmony' of the Copernican system, which recommended it to the astronomers of the day, might really be its constructive justification.

Chapters 8, 9 and 10 look at observations. As all observation is theory-laden, there is a question as to whether any constructive tree can ever terminate. The observations, on which it depends, require the truth of hypotheses that back them. Surely these hypotheses need to be justified in the same way that other auxiliaries do? I answer that so long as it is publicly ascertainable which way the observation came out in the actual observed case, that fact can serve as evidence by itself. Even someone who doubts the way in which the outcome is described (that is, the theory-laden statement of it), must acknowledge that it did come out that way, and not the reverse way. Our sensory organs did in fact make the detection, after all. So that fact in itself is significant, even if one doubts the theory with which some people load the outcome. Eventually, of course, it is very desirable that the theory in the background be independently confirmed. Chapter 9 looks at how to do this, and shows that here too we need to prohibit cycles when such independence is lacking.

Chapter 10 looks further into observations by asking why we should not only believe in what a theory says about the observable world, as Bas van Fraassen recommends. The answer is that even to differentiate the observable world requires that we make the best sense of the way our experiences are integrated into constructive justifications. Put differently, we sometimes trace a fault to an illusion or hallucination, or artifact. When we do so, we do so constructively. So the discrimination of what counts as the observable world requires that different experiences be integrated in a way that reaches common agreement. So, if one believes in the observable world, one agrees that different experiences must be integrated in this way. But if one agrees to that, then one discovers that the evidence for unobservable entities consists in just this kind of common integration among the observations. So one has to accept justifications for an unobservable world as well as for an observable world, and for the same reasons.

The final two chapters look at consequences and potential applications. Chapter 11 shows the sense in which constructivism gives a view of science that is independent of human values. Chapter 12 is more speculative, and gives some potential applications for the constructive point of view.

1 An Example of a Constructive Tree in Darwin

The most obvious reason to dismiss constructivism as a reply to Quine-Duhem is that scientists do not present constructive trees in their work. By the same token, though, mathematicians do not present derivations in predicate logic with identity in their reasoning. Rather, one can reconstruct the things they do say in terms of set theory and first-order, and sometimes higher-order, logic. Just as important, mathematicians reject reasoning that contradicts these inferences. If we find something similar in the things that constructivism says about what scientists say, and what they find objectionable, that is sufficient to show that the view is illuminating about the way natural science, and those who practice it, argue and reason.

A clear example of reasoning that is amenable to constructivism, too large to be useful, is Ernst Mach's *The Science of Mechanics* (1893). Mach shows systematically how outcomes of observation can be mustered into systematic justifications for a succession of increasingly informative hypotheses, with the result encompassing classical mechanics. Such cases are not at all common however, so it's better to begin with something more typical and less unwieldy.

In this chapter, we will look at an example from Darwin's *The Origin of Species* (1859 [1965]). Darwin replies to an objection to his theory here. If the Quine-Duhem hypothesis were true, all he would be required to do would be to identify some way to see matters that saves his viewpoint in the face of evidence that apparently contradicts it. But he does a great deal more than that. As constructivism says he must, he presents a rough and ready constructive tree to show that the auxiliaries he prefers are better justified than those of the alternative analysis. He goes even further than that, though. He uses constructive reasoning to give additional constructive justifications for an auxiliary that he's already supported constructively. Constructive trees can be added to buttress a hypothesis, as well as criticizing the auxiliaries of opponents. Both processes play a central role in doing what the Quine-Duhem hypothesis says cannot be done – using the outcomes of observations to identify the flawed auxiliary in the face of apparently contradictory evidence.

1.1 Isolated alpine populations have a recent common ancestor

Darwin was concerned to address the objection that alpine species could not possibly have spread between distant mountain ranges and to arctic environments because they could not survive in the intervening lowlands. How then could such species have descended from a single ancestral population? It is hardly possible for the same species to evolve simultaneously at disconnected areas (Darwin 1859 [1965], 359–360).

Darwin presents the following argument:

> So greatly has the climate of Europe changed, that in Northern Italy, gigantic moraines, left by old glaciers, are now clothed by vine and maize. Throughout a large part of the United States, erratic boulders, and rocks scored by drifted icebergs and coast-ice, plainly reveal a former cold period. (1859 [1965], 360)

He then uses the ice-ages to reconcile the consistency of dispersed identical alpine species with descent from a common ancestor:

> By the time that the cold had reached its maximum, we should have a uniform arctic fauna and flora, covering the central parts of Europe . . . As the warmth returned, the arctic forms would retreat northward, . . . And as the snow melted from the bases of the mountains, the arctic forms would seize on the cleared and thawed ground, always ascending higher and higher. Hence . . . the same arctic species . . . would be left isolated on distant mountain-summits . . . and in the arctic regions. (1859 [1965] 360–361)

Constructivism frames Darwin's argument as a tree (Fig. 1).

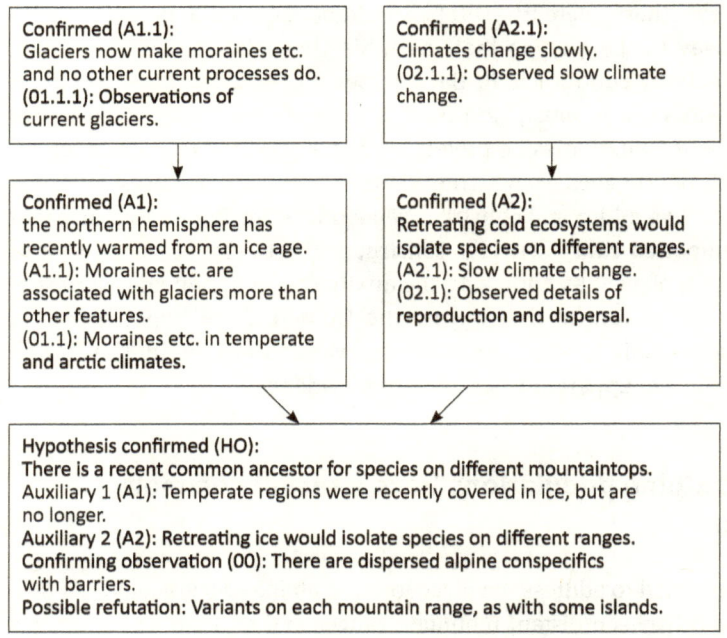

Figure 1: Isolated Mountain Species have a Common Ancestor.

1.1 Isolated alpine populations have a recent common ancestor — 23

This is a constructive tree.[3] The leaf nodes, at the top, are simply generalizations from the evidence, and do not depend upon additional auxiliary hypotheses. The hypotheses that are confirmed in the leaves then serve as auxiliaries in the justification in the succeeding node. The arcs, then, represent the relation between a parent node that *is a justification for* an auxiliary used in the child node. This illustrates the sense in which the target hypothesis has a justification that is *constructed* from the outcomes of observation. (One could reverse the arrows without changing constructivism materially. The important point is that the tree fills out one way in which our justification for believing the target hypothesis is substantiated by the fact that observations had one outcome rather than another.) Finally, in the root node at the bottom, Darwin confirms his target hypothesis.

Like Darwin, I have given a rather vague time-frame for the events in the tree. By 'slow climate change', for example, I only intend 'slow compared to the reproductive rate of the organisms' so that populations have time to move and are not wiped out by sudden dramatic changes. The time-frame may be vague, but there are clear cases on either side of the vagueness.

It is often easier to follow a constructive tree by displaying only the target hypothesis confirmed at each node. I call such trees *skeletons*. The skeleton of the above tree is illustrated in Fig. 2:

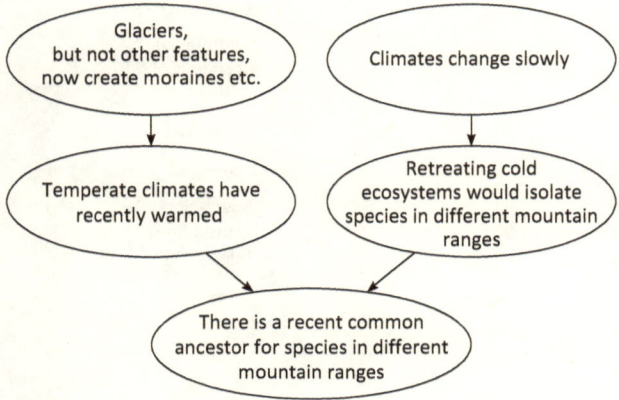

Figure 2: Skeleton of the Tree for the Common Ancestor.

---

3 A word on the numbering system for the nodes of the tree. I introduce the dot symbol as a diacritic, and justify the auxiliary at a new node indexed there. Suppose at node n I have three auxiliaries that are used to justify the target hypothesis, A_n, by using $\{O_n\}$. I name the auxiliaries $A_{n.1}$, $A_{n.2}$ and $A_{n.3}$ and justify them at nodes having those indexes. The exception is the root node where H_0 is justified. The constructive trees are intended to be traversed in a depth-first, post-order fashion.

24 — 1 An Example of a Constructive Tree in Darwin

Two constructive trees can be represented on a single skeleton. I will give another example made from materials in Darwin's *Origin*. The target hypothesis is that speciation occurs by descent with modification from a common ancestor. Darwin supported this with examples from adaptive radiation in islands, but he also used the modifications apparent from the fossil record to support the same idea. The two skeletons are separated in Fig. 3:

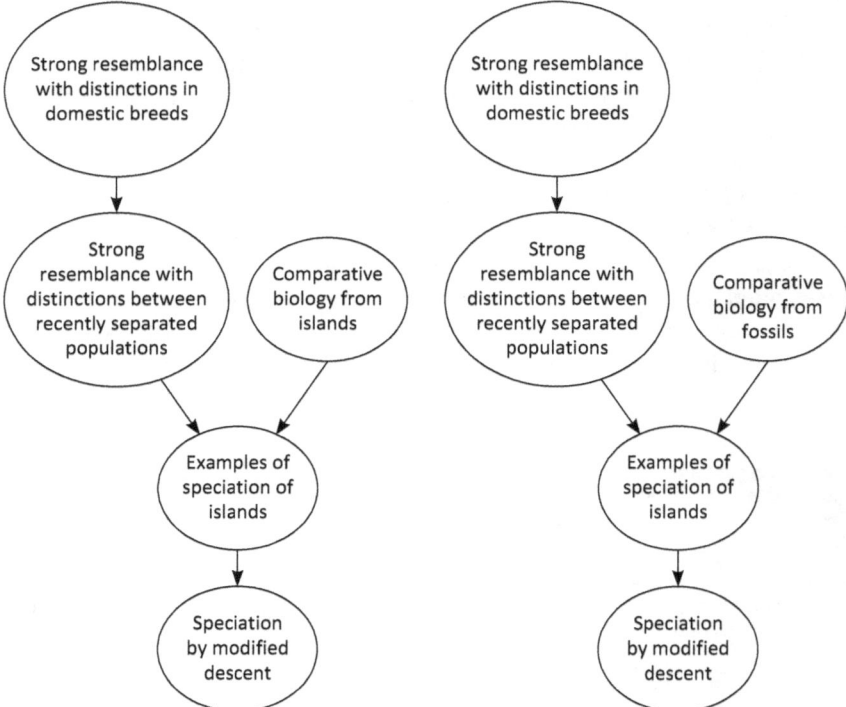

Figure 3: Separate Skeletons for Speciation.

But we can represent the two in terms of a single skeleton, as shown in Fig. 4.

One loses information when one moves from the full constructive tree to the skeleton. It isn't always possible to reconstruct the original constructive trees from their joint representation.

Skeletons are useful because they represent constructive trees more compactly. But they are also useful because they can trace a common dependency on

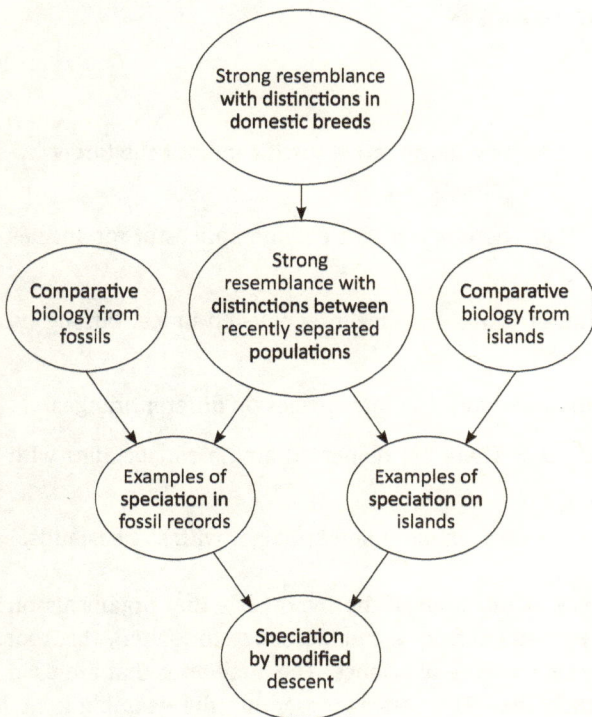

Figure 4: Single skeleton for Speciation.

a hypothesis that is shared by two (or more) constructive trees. In this case, for example, the two justifications both use the idea that one can identify populations with a recent common ancestor by looking for similarities and differences between them. That hypothesis has some evidence for it from domestic breeds of animals and plants. All the same, it would be nice to get more evidence, since the justifications from both fossils and islands depend upon it, so that both would be undermined if it proved false. If there had been strong independent evidence against the idea of speciation by descent from a common ancestor, the shared hypothesis would have been an obvious place to look for a flaw undermining both lines of support.

1.2 Explanation of the example

1.2.1 The root node

At the bottom of the constructive tree is the target justification of the target hypothesis, H_O:

> Hypothesis confirmed (H_O): There is a recent common ancestor for species on different mountaintops.
>
> Auxiliary1 (A_1): Temperate regions were recently covered in ice, but are no longer.
>
> Auxiliary 2 (A_2): Retreating ice would isolate species on different ranges.
>
> Confirming observation (O_O): There are dispersed alpine conspecifics with barriers.
>
> Possible refutation: variants on each mountain range, as with some islands.

This node of the tree offers a justification of the hypothesis that organisms on different mountain ranges descended from a common ancestor. Often, this root node is all we see when reading a piece of science. The auxiliaries that are used in the justification aren't defended. The person presenting the example is addressing a scientifically literate audience, who already know that certain hypotheses have been established. The presenter can select among these background hypotheses without having to defend them.

The target hypothesis is justified by citing observations together with the two auxiliary hypotheses. Those observations justify the target hypothesis if the auxiliaries are also justified. To repeat something I mentioned earlier: how exactly these justifications occur will vary depending up which theory of relative confirmation you prefer. Bayesianism will claim that making the observations in the root node increase the probability of the target hypothesis given the auxiliaries (see, for example, Howson and Urbach 1989; Strevens 2012; Maher 1993, 2004). Hypothetico-deductivism will urge instead that the observation is entailed by the hypothesis and the auxiliaries. Likelihood theories of relative confirmation will say that the observation is more probable given the hypothesis and the auxiliaries than it is given the denial of the hypothesis with those auxiliaries (Edwards 1969, 1972; Royall, 1997). Statistical theories of confirmation will say that the observation is very improbable if the target hypothesis is false, and probable if it is true (Mayo 1996).

Clark Glymour's bootstrap theory of confirmation and some versions of hypothetico-deductivism require that we cite potentially refuting evidence, as well as the confirming evidence which we observed (Glymour 1980; Hempel 1966, 8). I think it's easy to see how to produce this in this example. Popper did not propose a detailed theory of confirmation, but his demand for falsifiable hypotheses clearly fits with this idea (1959, 1963, 33–59).

1.2.2 Confirming the auxiliaries

There are two auxiliaries, each supported at a node. The recent warming trend is supported thus:

> Confirmed (A_1): the northern hemisphere has recently warmed from an ice age
>
> Auxiliary1.1($A_{1,1}$) Moraines etc. are associated with glaciers more than other features.
>
> Data ($O_{1,1}$): Moraines etc. in temperate climates and the arctic.
>
> Possible conflict: no such evidence, (as in the tropics).

There's also a node justifying the isolation of the different ecosystems as the climate changes:

> Confirmed (A_2): retreating cold ecosystems would Isolate species on different ranges.
>
> Auxiliary2.1 ($A_{2,1}$): Climate change is slow.
>
> Data ($O_{2,1}$): Observed slow climate change over historical time.

The idea is that these two nodes provide a justification for these two auxiliaries that is independent of the justification that the root node provides for the target hypothesis (that the dispersed individuals share a common-ancestor).

Thus, the root node provides a justification for the common-ancestor hypothesis, the hypothesis that the whole tree justifies. Such justifications normally rely on auxiliaries, and there must be at least one node connected to the root that justifies each of the auxiliaries it uses. This obviously begins a chain of justifications, since the justification of one auxiliary normally requires other auxiliaries. The chain of justifications is a path through the tree, a branch. The hypothesis that is justified at each node gets used as an auxiliary in the justification offered for a different hypothesis at the next node. So long as each of these paths begins with a leaf, no hypothesis that is needed for the justification offered in the root node goes without some evidence in its favor.

One can follow a chain of justifications in this tree then. The recent-common-ancestor hypothesis is justified relative to the auxiliary that the temperate zones of the earth have recently undergone an ice age. The hypothesis that they've undergone an ice age is confirmed by the observation of moraines and other features of glaciation, relative to the auxiliary that glaciers, and not other things, cause these features. And that auxiliary is confirmed in turn by observing that glaciers that currently exist are producing these features, and no other processes are currently producing them.

One could object, at this juncture, that Darwin has not explained how slow climate change would lead to isolation of descendant populations in suitable ecosystems. He would need an additional auxiliary, something like:

A3: If a change in habitable ecosystems is slow (compared to the reproductive rate of individuals), then individuals in the later ecosystems have a common ancestor in the earlier one.

It's easy to see why Darwin wouldn't include this auxiliary, and the evidence for it. It is simply obvious to all readers that there is evidence for it in, for example, the spread of weeds from the wilderness as agriculture is introduced, or the invasion of new ecosystems by introduced species, or the succession of plants after a forest fire. Everyone knows that organisms are dispersed to suitable new areas as an ecosystem moves slowly in space and time. It's a feature of ordinary reproductive mechanisms and it would insult the intelligence of the reader to discuss it.

1.2.3 The leaves

The leaf nodes are the only nodes that offer justifications without using auxiliaries. Since there are no auxiliaries, there is no need to justify anything further, and the path begins (or terminates, depending on which way you look at it).

One leaf gives us the evidence for recent glaciation:

Confirmed ($A_{1.1}$): Glaciers now make moraines etc. and no other current processes do.

Data: ($O_{1.1.1}$) Observations of current glaciers.

Possible Conflict: No moraines in current glaciers. Moraines in tropical climates.

The other argues that climate change is slow:

> Confirmed ($A_{2,1}$): climates change slowly.
>
> Data ($O_{2,1,1}$): Observed climate change is slow.
>
> Possible conflict with observed fast changing climates.

To repeat, all that is meant by the hypothesis here ($A_{2,1}$) is that the climate change is slow *compared to the reproductive rate of the species involved*, so that a changing climate provides time to gradually move location.

In this example, and in the other examples in this book, the leaf nodes are generalizations from the data. This might be straightforward induction, or support for a hypothesis involving probability from a sample of data. (In Glymour's bootstrap theory of confirmation another kind of leaf node is possible, one in which a hypothesis is confirmed using itself as the only auxiliary: 1980, 110–133.)

The leaf nodes need not be at all secure, as foundationalism would presumably demand. Nor are they some privileged kind of justification that is immune from doubt, or isn't part of some theory. Every node provides a justification from the data, and the hypotheses that are justified at the leaf nodes can be the target hypothesis of another tree. Even an observation that is used at the leaf of one tree can be the target hypothesis of another. Indeed, I will try to show that that is one way to track down unreliable observations.

We can put constructive trees to at least three uses, as follows.

1) Relative to real confirmation. Constructive trees show one way in which a target justification can be completed. The point of showing that is to show that we are entitled to believe the auxiliaries for reasons that do not depend upon either the evidence cited in the target justification, or the target hypothesis. The point of that, in turn, is to show that we are entitled to regard the evidence as bearing upon the target hypothesis. Some might prefer to regard favorable evidence for a target as showing that some auxiliary we have used is false (Quine 1953 [1980], 43). While a constructive tree has not decisively shown that that cannot be done, it has shown at least one reason for not regarding the evidence that way. Some of the things we have observed show that the auxiliaries are true, whether we believe or disbelieve the other hypotheses in the target justification. So we have real confirmation, if only weakly and defeasibly. (I have written of belief and disbelief here, although some theories of relative confirmation do not focus on doxastic issues. I will continue to do so to keep matters simple.)

One might put the point in terms of a possible history of justifications. By giving a constructive tree, one shows that it is at least possible for the hypotheses at the leaves to have been justified first. Some reasoner might have accepted this justification without having any idea whether the target hypothesis was true or whether the observations of the target justification would be made. Then we go through the tree, showing the reasoner eventually winding up at the point that we want to show is a possible epistemic situation. That is, eventually our reasoner might have a justification for the auxiliaries in the target justification, but be completely in the dark about whether the confirming or refuting observations will be made, and wholly agnostic about whether the target hypothesis is true. Witnessing the confirming observations justifies believing the target hypothesis. So making those observations should indeed raise the credibility of the hypothesis, because we know of one set of evidence that justifies us in believing the auxiliaries, and those auxiliaries, when the new evidence is added in, justify us in moving from agnosticism to belief in the target hypothesis.

So long as this kind of argument will go through, that is all we need for the transition from relative to real confirmation. So constructivism gives us a model of independent confirmation that we can use to move from relative to real confirmation. But there are other uses for constructive trees in addition.

2) Replying to objections to a theory. We have seen that unlike foundationalist theories, the leaves of a constructive tree need not be at all secure. The observations are not privileged sense-data but are events that are predicted, or forbidden, according to some theory. So constructive trees need not be at all strong. To be sure, the tree must provide a justification for the target hypothesis, but there could easily be *other* evidence, or other hypotheses, that are much better justified and which deny the target. We can challenge one constructive tree with another. Indeed, that was why Darwin first presented this example.

Darwin was responding to a potential criticism of his view in the *Origin*, from the simple tree as depicted in Fig. 5:

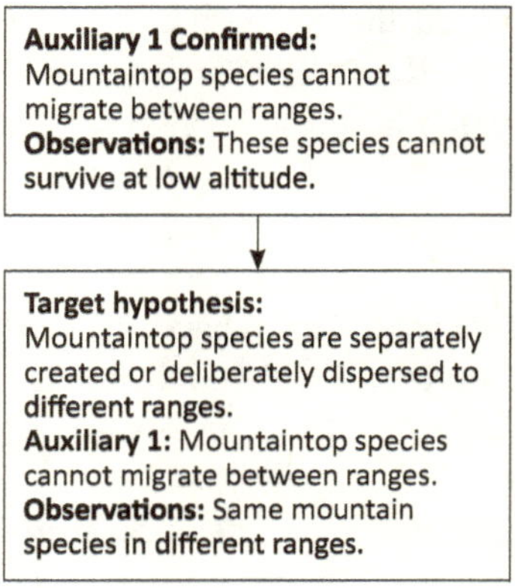

Auxiliary 1 Confirmed:
Mountaintop species cannot migrate between ranges.
Observations: These species cannot survive at low altitude.

↓

Target hypothesis:
Mountaintop species are separately created or deliberately dispersed to different ranges.
Auxiliary 1: Mountaintop species cannot migrate between ranges.
Observations: Same mountain species in different ranges.

Figure 5: An objection to Darwin's justification.

Darwin is obviously challenging the auxiliary hypothesis that there is no way for an alpine species to migrate across the lowlands.

A constructive tree shows how the auxiliaries we use at the root node possess *some* justification that we can be sure is secure whether or not the target hypothesis is true. If only very weak examples are forthcoming, it will obviously look feasible to challenge the auxiliary in order to evade the evidence for the target. If we can produce very strong support for the auxiliary, that will count against such a strategy.

One important advantage to constructive trees is that in real science, we cannot just challenge an auxiliary for free. Darwin didn't say that conspecifics that are isolated in different ecosystems *might have had* a recent common ancestor because the environment *might have been colder*. He gave evidence to support the view that these hypotheses were better justified than the idea that the species had been isolated for an evolutionarily significant period of time. Darwin's imaginary opponent gave a reason for believing his auxiliary, and Darwin was obliged to provide a better reason for thinking that it was mistaken.

This is a different picture of science from the one we have inherited from Quine and Duhem. There, the emphasis was on the fact that the auxiliary can be challenged given the data at the root node. In the Introduction I cited many quotations stating that we could hold onto any hypothesis in spite of any evidence, or refuse to accept any hypothesis in the face of any evidence, by altering the auxiliary hypotheses. Yet there can be additional evidence that confirms the auxiliary no matter which way the data at the root node come out. Darwin presented this additional evidence in this example.

Any normal scientist can do the same when he depends upon an auxiliary, and very frequently does so explicitly. Working scientists check auxiliaries when they try to set up independent experimental tests. Students ask their professors why they can depend upon auxiliary hypotheses without assuming what is supposed to be demonstrated, and professors answer them. Scientists criticize each other for using insecure or question-begging auxiliaries, and cite evidence to defend them as independently justified. Constructivism makes intuitive sense of the things we see in scientific practice on the face of it. The question is whether it can be spelled out successfully.

3) Buttressing support for a hypothesis with an independent constructive tree. One way in which we can strengthen a target hypothesis we have constructively confirmed is to present additional trees which also support it. It is open to the constructive theory of confirmation that the combined support of many very weak sources of support for a single hypothesis could together provide a very

strong justification for it. A close analogy might be simultaneous signals from many independent detectors, each of which is unreliable much of the time. Each boy might cry "wolf" to amuse himself sometimes. But if they are independent, and all cry "wolf" together, then the hypothesis that there really is a wolf may be more secure than the veracity of any one of them.

We can produce independent constructive trees for auxiliaries in a tree as well. An auxiliary hypothesis might be supported, not at a single incident node, but at many of them. In the example of the separated alpine conspecifics, Darwin extends the justification for using one of his auxiliaries by adding additional sub-trees. The leaf node confirming 2.1 justifies the hypothesis that climate change is slow compared to the reproductive rate of species by looking at the observed rate of climate change. Now how many examples of observed climate change did Darwin have? Only one, surely, the one of historical time, which is hardly a large N for the father of modern biology.

Darwin did, though, present other cases of stable climates in the *Origin*. He refers, for example, to very thick deposits of fossil shellfish of various kinds around the world (1859 (1964), 282–292). These are so thick that many generations of animals must compose them. So it is reasonable to extend this branch a little by adding a sub-tree (see Fig. 6).

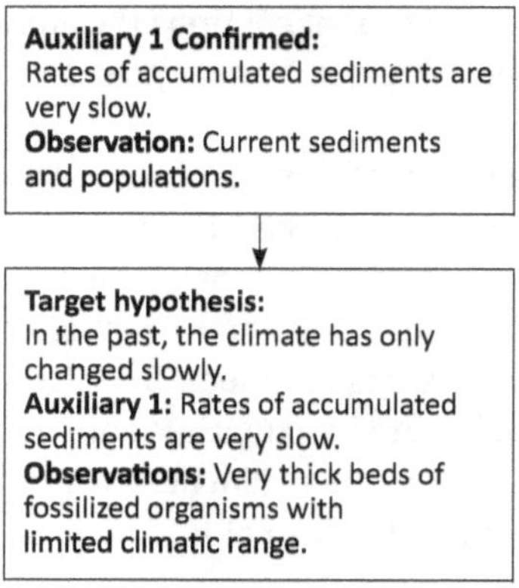

Auxiliary 1 Confirmed:
Rates of accumulated sediments are very slow.
Observation: Current sediments and populations.

↓

Target hypothesis:
In the past, the climate has only changed slowly.
Auxiliary 1: Rates of accumulated sediments are very slow.
Observations: Very thick beds of fossilized organisms with limited climatic range.

Figure 6: Independent Justification for Slow Climate Change.

A poor constructive tree, which provides very little justification for some auxiliary in the target justification, can be buttressed with additional evidence. To abandon the auxiliary in the root node will require challenges to more and more justifications for it. If these come from a wide range of topics, that will make it harder to eliminate them all.

This is one reason for the importance of the variety of evidence in justifying a hypothesis (Glymour 1980, 139–142; Lloyd 1983 [1994], 149–152). Glymour wrote,

> if a hypothesis is confirmed by observations . . . using another hypothesis . . . then it is always possible that the agreement between hypothesis and evidence is spurious. The only means available for guarding against such errors is to have a variety of evidence, so that as many hypotheses as possible are tested in as many different ways as possible. What makes one way of testing relevantly different from another is that the hypotheses used in one . . . are different from the hypotheses used in the other. (1980, 140)

Constructivism provides a way to elaborate this thought. The different ways of testing a single hypothesis are different constructive trees, each of which has that hypothesis as its target, but which do not share many auxiliary hypotheses. Of course, a hypothesis that serves as an auxiliary (or an auxiliary to testing an auxiliary, and so on) can be taken in a different context as a target hypothesis.

This example, and others like it, will probably raise many questions. So let's look at a few and their answers.

1.3 Can this example be made complete?

I have granted that Darwin omitted to mention and defend one of the auxiliaries he used:

A3: If a change in habitable ecosystems is slow (compared to the reproductive rate of individuals), then individuals in the later ecosystems have a common ancestor in the earlier one.

It could be argued that the process of adding unmentioned auxiliaries cannot be brought to an end. Darwin's example mentions time and space, for example. Someone could object that it is an auxiliary that time is one-dimensional, and that different spatial points are related by some distance. Constructivism is doomed if this sort of extension goes on indefinitely. It should be possible to construe Darwin's example as having only a finite number of auxiliaries, otherwise they cannot all be supported in a finite tree.

To show that the example can be brought to an end, I will now show one way of doing so. There are three auxiliary hypotheses in the target justification. Then, along with the observations, we can represent the target hypothesis as a first-order logical consequence of these auxiliaries. I do not think this is the only way to present the justification Darwin gave in the passages above. But it is one way to do so, and it makes the hypotheses explicit and uses only first-order predicate logic for the inferences.

We have three auxiliaries:

A1: Any two European locations have recently (that is, less than 20,000 years ago) undergone a major climate warming.
A2: When two locations undergo major climate changes, the changes in the ecosystems are slow (with respect to the reproductive rate of a species).
A3: Whenever movements in ecosystems are slow, species will migrate to the new ecosystem.

The next step is to represent these as sentences of first order logic.

The vocabulary is as follows:

$M(x,y,t1,t2)$: There's a major climate warming between t1 and t2 for locations x, y.
$C(x,y,t1,t2)$: The ecosystem of x at t1, becomes y at t2 slowly.
$S(x,t)$: some example of a species (for example, the alpine marmot) inhabits x at t.
$E(x)$: x is a location in Europe.
$CommA(y,t1,x,t2)$: Individuals living at x at t2 have common ancestors in y at t1.

Formulating the auxiliaries:

A1: $\forall x \, \forall y \, ((E(x) \wedge E(y)) \rightarrow (\exists t \, (Recent(t) \wedge M(x, y, t, n))))$
where n refers to today.

A2: $\forall x \, \forall y \, \forall t1 \, \forall t2 \, (((Alpine(x, n) \wedge Lowland(y, n) \wedge M(x, y, t1, t2)) \rightarrow C(x, y, t1, t2))$

A3: $\forall x \, \forall y \, \forall t1 \, \forall t2 \, ((C(x, y, t1, t2) \wedge S(y, t2)) \rightarrow S(x, t1))$

We need a hypothesis to take care of the idea of common ancestor:

CA: $\forall x \forall y \forall t1 \forall t2 \, ((C(x, y, t1, t2) \wedge S(y, t1) \wedge S(x, t2)) \rightarrow CommA(y, t1, x, t2))$

I have noted that Darwin could reasonably claim we have evidence of this from the same phenomena as give us evidence for A3.

Finally, let a and b be two now widely separated alpine environments, say in the Alps and the Carpathians. The conclusion Darwin wants is:

H: $\exists x \, \exists t \, (CommA(a, n, x, t) \wedge CommA(b, n, x, t) \wedge Recent(t))$.

Find a lowland environment, *c midway between a and b*. By A1, there was a recent time when *c* was as cold as the Alps are now. By A2, *c* warmed slowly. And by A3, species in the lowlands move to the Alps. By exactly the same logic, they also moved from there to their current position in the Carpathians. Which gets us the conclusion.

This is no doubt a very inept and flat-footed representation of Darwin's reasoning. I do not deny that the objection is correct in thinking that one *can* come up with a reconstrual of the example where one demands, for example, more detail on what makes something recent, or on what hypotheses spatial and temporal relations satisfy that underwrites the inferences that this reconstrual takes for granted. What I deny is that one *has* to proceed in a way that makes this process interminable. One can understand Darwin, and practice science oneself, without doing so. The deductive justification here could be replaced with inductive ones according to some theory of relative confirmation. Sometimes we draw conclusions even if we know that the conclusion is fallible, and that future investigation might show we are mistaken. Science would be impossible if this were not so, for we would never be able to draw an empirical conclusion.

1.4 Why are examples of a constructive tree not more common?

Scientists typically present justifications to other scientists. Both parties to the debate are aware of the evidence available to each. Both parties have also been through a process of science education, which has informed them of what has and what hasn't been observed, and how various hypotheses have been justified from these observations. Since both of them know this, it's irrelevant to the purposes of their engagement. The background isn't filled in, not because it isn't needed for the justification, but because both of them are working in a context where they are both aware that they both already know it. What is of interest is the new observations one is communicating to the other, and the reasoning concerning the way in which those observations bear upon some target hypothesis.

The issue, then, of when we should expect to see constructive trees – or parts of them – and when not, is an issue of the purposes the presenter and recipient of a justification have when they communicate. It is an issue of how much a presenter needs to say or write to communicate one to his audience. In the philosophy of language, Paul Grice argued that there was a maxim of Quantity that guided conversations. People sought to provide as much information as is required for the purposes of the exchange, but no more than that (Grice 1989, 26).

It is easy to see how this maxim, applied to the communication of justifications, would discourage two knowledgeable people from communicating everything that is necessary for a justification to work. The presenter is trying to be informative *to the recipient*, and so doesn't present information that the recipient already possesses. Because of a shared background, both participants are aware that the auxiliaries are supported by the evidence, and it becomes redundant to present them. One doesn't waste words; it's simply pointless, and boring for the recipient, to go through things that he or she already knows.

But if constructivism is going to take this line, then it has to make good on the assumptions on which it depends. In particular, constructivism has got to somehow show that it is reasonable to view the processes scientists go through in their education as providing them with the wherewithal to fill in the justification for the auxiliaries in a way that is independent of the target hypothesis in typical justifications. And in particular, the constructivist is going to have to make out the sense in which two scientists could, if they wanted to, eventually trace the independent justifications for the auxiliaries back to observable outcomes that either we have actually observed, or at least have very good reasons to think that we would observe if we were to engage in certain experiments. When you think about examples such as the justification for the existence of the Higgs boson, or hypotheses concerning the processes involved in a supernova explosion, this is daunting.

A great deal of this book is devoted to showing that such a thing can at least get started. That is what I see as the central challenge. Once one can see how very basic reasoning gets done constructively, the practice of science education helps with the rest. For that process takes these basic starting-points and builds upon them.

1.5 When should we expect to see scientists (and others) giving constructive trees, or parts of them?

We should expect this when the conditions cited in the last section break down. So we should expect a constructive tree, or a part of one, under circumstances like this:
1. The auxiliary may be controversial. A scientist might be responding to an objection that attacks or questions the auxiliaries he or she is using.
2. It might be questionable whether the auxiliary can really be justified independently of the target hypothesis. In particular, it might be doubtful whether it is possible to trace the justification back to generalizations from the observations in a non-circular way (or, indeed, at all).

3. The scientist might be addressing a less well-informed audience, such as colleagues not in the field. The scientist needn't fill out the whole of the independent justification here, but should at least give them enough information to know how to find out for themselves.
4. There might be a pedagogical purpose. At least some textbooks, (for example Ernst Mach's *The Science of Mechanics* (Mach 1883)) read as if they are trying to account for the whole of our knowledge in some restricted area by arguing from the behavior of objects that students could observe with their unassisted senses.
5. A scientist might just be trying to be complete and convincing, as Darwin probably was in this example.
6. A scientist might be trying to identify where some experiment has gone wrong, or which element of some set of hypotheses is responsible for a conflict with the data.

In the case of most of these purposes, the scientist presenting the target justification doesn't need to be exhaustive in going through each path of justifications in the tree. He or she only needs to follow them down as far as he needs to address the puzzlement or objection. Constructivism predicts that scientists and others will pursue the justification only as far as is necessary to address the motive. We need to outline as much as our audience needs. But we need not often proceed very far to address this problem.

We at least try, within the constraints of time and budget, to explain why students ought to accept the assumptions of our demonstrations (to take one among the uses of the trees). What is doubtful is not that we do these things *to some degree*. What's doubtful is that we are even able to do them to the degree that the constructivist says we must be able to. Constructivism demands that we ought to be able to pursue the chains of justifications all the way to observations, even if we do not usually do so because of constraints of time, or money, or interest. So we return to the question that we have already visited repeatedly. Is it even possible to do this?

In the next few chapters I will try to show the following: Justifications that can be made into constructive trees are ubiquitous in science. Justifications that cannot are unusual, and suspect. I will also try to persuade you that when scientists write and speak of confirming something independently of some hypothesis, constructivism makes sense of what they do. And I will offer a theory of the way in which constructivist justifications track down errors. These are the main ways I will argue that the terminus of unobjectionable justifications lies in the outcomes of observation.

So in this chapter we have, as I see it, an example of a constructive tree from an actual case in the literature. I do not think incomplete examples of constructive trees are unusual elsewhere. Examples that are even as complete as Darwin's are unusual though. Constructive trees are often tedious. They depend upon theory-laden outcomes of observation, they give a lot of credit to induction, and they aren't infallible. But having said all that, they do show how some hypotheses can be justified independently of other hypotheses, and they give us a way to get real confirmation. Nothing else does so, at least so far, and we do need to be able to do these things to practice science.

2 The Meaning of Independent Confirmation

According to constructivism, each auxiliary in a target justification has to be justified independently of that target. So we get a new justification, which focuses on each auxiliary, and a justification for the auxiliaries in *that*, and so on until we dead-end in a justification that appeals to no further auxiliaries – that is, a justification by induction from outcomes of observation.

This is a recursive structure. Any justification by induction using no auxiliaries is a constructive tree, and given a set of constructive trees justifying a set of sentences $\{A\}$, the tree formed by independently confirming H from $\{E\}$ independently of the justifications for $\{A\}$ is also a constructive tree when it forms the root of the trees justifying $\{A\}$.

But what does independent justification mean? What is it for a justification to be independent of another justification, or independent of a hypothesis? This chapter spells this out using Bayesianism. It gives two conditions which, when both fulfilled, give a sufficient condition for a justification of $\{A\}$ to be independent of the justification of H by $\{E\}$. Perhaps these are too strong, and disallow desirable justifications – although I know of no examples here. I argue that it disallows the counterintuitive examples of justification in the literature.

There is a particularly important consequence of these conditions. No constructive tree can contain its own target hypothesis as one of the hypotheses on which its justification ultimately depends. So constructivism must prohibit any *cycle* of justifications. These are cases where both some hypothesis $H1$ is justified from the evidence using only trees containing some hypothesis $H2$ and $H2$ is justified using only trees containing $H1$. In such a case we would possess no constructive trees justifying either hypothesis, so there is no constructive justification for either. The idea is extended to cycles containing more hypotheses in the obvious way so that it is a cycle if $H1$ requires $H2$ which requires $H3$ which requires $H1$.

Under the influence of Quine-Duhem, one would think that hypothesis independence must rule out a legitimate form of justification. For if the unit of empirical significance is the whole of science, then the reason we believe any scientific theory must be the fact that it hangs together as a whole, both with itself and other scientific theories, in the light of our observations. So if we try to chase down all the justifications for the auxiliaries we use in some experiment confirming H, we are almost bound eventually to come to an experiment that depends upon H. According to Quine-Duhem, this is to be expected and is a perfectly reasonable chain of justifications for the auxiliary. Constructivism has to deny this.

Cycles are important because they allow constructivism to account for a particularly influential kind of objection to justification in the history of the philosophy of science. This chapter closes by looking at an example that has been subject to this kind of objection, Newton's original proposal for absolute space.

2.1 Observation independence and hypothesis independence

There are two conditions that, according to constructivism, are jointly sufficient for real confirmation:

> **Observation independence**: a constructive tree is observation independent (of an observation in the target justification) if and only if using the constructive tree to justify an auxiliary doesn't require belief in *those outcomes of observation that are cited in the target justification.*
>
> **Hypothesis independence**: a constructive tree is hypothesis independent (of the target hypothesis) if and only if using the constructive tree to justify an auxiliary doesn't require belief in *that hypothesis which is confirmed in the target justification.*

Why think that these conditions are individually needed and jointly sufficient for the kind of independence that constructivism ought to seek?

One reason is that real confirmation is suppressed to the degree that either condition is violated. Consider first an investigator who violates observation independence. Such an investigator ought to be in a position of having a justification for all the elements of $\{A\}$, but unsure whether or not H. Upon observing that $\{E\}$, H then gets additional justification. But if justifying $\{A\}$ requires believing that $\{E\}$, and $\{E\}$ justifies H, the investigator can hardly be in much doubt about whether or not he's going to observe $\{E\}$, once he's convinced of $\{A\}$. H will already have received the benefit, even if we never check up on $\{E\}$ by observation. Even if, upon inspection, things do not go as expected, and $\neg E$ gets observed instead of $\{E\}$, the result is only an incoherent set of beliefs, not the refutation of H in the light of $\{A\}$. Intuitively, the justification of the As and the justification of H by $\{E\}$ interact too much; they are not independent.

Now consider violations of hypothesis independence. Upon observing those outcomes that justify $\{A\}$, the investigator is required to believe that H. Well then, observing outcomes $\{E\}$ can hardly add very much confidence to that hypothesis. The As have not been confirmed independently of H. And just as in the case of observation independence, even if we in fact observe some $\neg E$, instead of $\{E\}$, the result is an incoherent set of beliefs, not a refutation of H in the light of $\{A\}$. We know something must be wrong somewhere – either both H and some A are wrong, or some justifications for some A has gone wrong, or the making of the observation is somehow misleading. We do not have a justified background that has increased the credibility of H over $\neg H$ on observing $\{E\}$ as opposed to some $\neg E$, which is what real confirmation requires. We have instead The Great Muddle that Quine-Duhem says is inevitable because independent justification is really impossible.

The idea that each condition is necessary for real justification is reinforced by examples in the literature, mostly due to David Christensen. These are examples that are highly counterintuitive as pieces of justification, but which violate either hypothesis independence or observation independence. They were all presented originally as counterexamples to Clark Glymour's bootstrap theory of confirmation, though they apply to other theories of relative confirmation too. Because they are strongly counterintuitive, they wouldn't be accepted by anyone as instances of real confirmation. And each of them violates either observation independence or hypothesis independence (David Christensen 1983, 1990, 1997; Aaron Edidin 1981; Peter Achinstein 1983).[4]

2.2 Violations of observation independence: Examples

Consider the following example from Christensen, which I will call 'ravenfeather' (1983, 479):

Hypothesis H1: $\forall x(Rx \rightarrow Bx)$ ("All ravens are black")

Auxiliary, A1: $\forall x(Rx \rightarrow (Bx \leftrightarrow Fx))$ ("All Ravens (are black if and only if they have asymmetric primary flight feathers)")

Evidence E: $\{Ra, Fa\}$

Possible Counterevidence E': $\{Ra, \neg Fa\}$

---

4 Christensen (1983, 1990, 1997) provided counterexamples to Glymour's bootstrap theory of confirmation. Glymour (1983) and Glymour and Earman (1988) initially proposed replies, to which Christen responded (1990). Christensen showed how instances of the counterexamples under different interpretations of the non-logical vocabulary changed from being counterexamples to being intuitively plausible (1990). John Earman argued that Christensen's work successfully showed that bootstrapping should be abandoned, and suggested that Bayesianism was a more promising line for work on confirmation theory to pursue (1992, 73). Constructivism was first suggested in the literature on bootstrapping by Jan Zytkow (1986, 108). Sam Mitchell (1995) took up his suggestion.

While the example was originally proposed in opposition to Glymour's bootstrap theory, it presents problems for other theories of confirmation too. If A1 is the background knowledge for Bayesianism, for example, the posterior probability that all ravens are black goes up if we observe that a raven has asymmetric primary flight feathers, at least if one is in any doubt about whether or not the flight feathers are symmetric. That seems at least odd.

Christensen gives other examples in which *A1* is replaced by other auxiliaries which have the same logical form, but which give rise to examples which do not seem at all odd:

A2: AIDS patients have antibodies to HIV if and only if they have been infected with HIV.

A3: Halmasauruses have fractured heelbones if and only if they were avid jumpers.

It doesn't seem odd to confirm that AIDS patients have HIV by discovering cases with AIDS who have the antibodies. Nor does it seem odd to justify the claim that Halmasauruses were avid jumpers by finding fossils with fractured heel bones. Why does ravenfeather look wrong when these cases look right?

In proposing examples and counterexamples to a theory of confirmation, we are asking whether the edicts of the confirmation theory about whether or not justifications are legitimate match our intuitions. In this case, we can see no way that a real person would use the suggested background knowledge (that all ravens are black if and only if they have asymmetric primaries) to confirm the target claim (that all ravens are black). Instead, the target claim would be a matter of direct observation to virtually any real person suitably situated.

The obvious way to confirm the auxiliary that all ravens are black if and only if they have asymmetric primaries is to look at a lot of ravens. But, if one does this, one automatically sees that they are all black. The background evidence for the auxiliary requires that we observe colors when we observe the shape of the feathers. So the background evidence shows that one cannot gain some instance of the target hypothesis without at least as much getting an instance of the auxiliary. The auxiliary cannot be confirmed in a way that is observationally independent.

Contrast this to the case of the Halmasaurus. No one functions normally in society without knowing that words ending in "-saurus" refer to dinosaurs, that they are extinct, and that we know of them through fossilized remains of their skeletons. So it is inevitable that we observe the heelbone of one without observing the living creature itself. Also, a fractured heelbone seems like a reasonable symptom that might lead a knowledgeable person to speculate that the animal jumped a lot.

It is a reasonable supposition that creatures who jump a lot might fracture their heelbones. We don't, when we observe a Halmasaurus skeleton, automatically observe its status as a hurdler. So we can all see how the observation of a fractured fossil heelbone can be part of a justification that the halmasaurus was an avid jumper. The example is intuitive because even if we haven't made the observations in detail, other confirmed hypotheses about the world do not make it improbable to us.

If this answer is correct, then changing the example so that the fossilized remains show something that isn't intuitively connected with jumping ought to make the example counterintuitive. I think this prediction comes out correctly. Replace the fractured heelbone with a fractured eardrum and the example looks silly. Now, the hypotheses upon which the example depends make it very improbable to us. Fractured eardrums seem intuitively to be a poor indication of whether an animal jumps.

The AIDS example is intuitively appealing for similar reasons. We are all able to see AIDS patients directly, but we are aware that both the virus and the antibodies require detection that is more elaborate. And if you have enough biology to know what an antibody is, you probably have enough to know that it is often easier to detect them than it is to detect the specific pathogens themselves. (Even if you do not know this, the idea that doctors might test for a disease by looking for something they call 'an antibody to' the pathogen is quite reasonable.)

You can test the suggestion in two ways. First, by replacing '. . . has antibodies to HIV' with something that is intuitively unconnected to AIDS, like having a vowel in your initials. The result should be, and is, a counterintuitive example. We do not look at the initial letters of patients' names to discover whether or not they are suffering from AIDS, and it would be silly to think that a patient has AIDS if and only if his or her initials contain a vowel. If the auxiliary looks manifestly false to the evidence, the example fails.

Second, you can try to make the AIDS case similar to the ravenfeather case. This is a little tricky, as AIDS patients do not have self-evident properties in virtue of having AIDS (particularly now, when they can live almost asymptomatically). A possible case is the property of having an illness. So we get 'All aids patients (have an illness if and only if they're infected with HIV)'. Anyone who doubts that HIV is the virus that causes AIDS won't, I think, allow this to function as an auxiliary. That person certainly isn't going to regard the observation that someone has AIDS, and is also ill, as evidence that all AIDS patients are infected with HIV.

I have proposed that the ravenfeather example looks counterintuitive because of the role that color perception plays in our detection of ordinary middle-sized objects like ravens and asymmetry. So again we can predict an "observation". The justification ought to stop looking so counterintuitive if the feather structure were a plausible indirect method for detecting the color of the bird. That, after all, is why I allege that the AIDS and Halmasaurus cases look intuitively correct.

I think this is borne out too. Every bird species that is not flightless has asymmetric primaries. Suppose, however, that the shape of their primary flight feathers reliably indicates the colors of birds. Red birds have one shape, yellow birds another, and so on. And now suppose you are blind. It might be an engaging party trick to be able to tell the kind and color of various birds that are handed to you, by feeling for the shape of their body and flight feathers. If you were avuncular, you might puzzle children as to how you were able to "see" the colors.

Alternatively, supposing that there is some feature of the structure of a feather that indicated what color it was, I think we can make the example intuitive in contexts in which we cannot observe the color of the feather. Archaeopteryx was discovered with the feather structure fossilized along with its bones, so we have no idea what color the original bird was. An established correlation in living birds between feather structure and color would certainly give evidence. If we discovered a fossilized raven with the feather structure intact, one might infer from this that ancient ravens, as well as modern ones, were black.

2.3 Violations of hypothesis independence: Examples

The next example, by Edidin (1981), takes a little setting up. Take a case where constructive justification is clearly possible. Consider the following theory (T1):

$$H1: \forall x(Sx \leftrightarrow Tx)$$

$$H2: \forall x(Ax \leftrightarrow Sx)$$

$$H3: \forall y(By \leftrightarrow Ty)$$

I'll suppose objects bearing every property do so observably. I'll also suppose that we have got lots of instances of H2 and H3, so that they're secure. Suppose it's very expensive and inconvenient to get oneself into a situation where one observes S-ness and T-ness for objects, but easy to observe the other properties. It seems perfectly reasonable to confirm H1 by observing that As and Bs are inevitably associated. Had we observed an A without and accompanying B, the hypothesis would have been refuted.

(For example: we observe that oxpeckers fidget if and only if they carry small ticks of some kind. Rhinos are lethargic if and only if they carry a larger version of the same kind of tick. H1 might be the result of the view that the tick moves from oxpeckers to rhinos as it matures. Surely, with this background, we

2.3 Violations of hypothesis independence: Examples — 45

could confirm H1 by observing that fidgeting oxpeckers are associated with lethargic rhinos, without having to catch and inspect the creatures.)

On the other hand, it seems utterly empty to confirm H1 of this theory (T2):

H1: $\forall x(Sx \leftrightarrow Tx)$

H2: $\forall x(Ax \leftrightarrow Sx)$

H3': $\forall y(By \leftrightarrow Sy)$

by, once again, observing that *A*s and *B*s are regularly correlated.

Well, why is this absurd? After all, the observations we are using – that *A*-ness and *B*-ness are invariably accompanied by *S*-ness, and that *S*-things are also *T*-things – are obviously consequences of T2. Under simple versions of hypothetico-deductivism, the confirmation is apparently analogous to the confirmation of *H1* in T1. T1 and T2 are equivalent theories under both semantic and syntactic views of the nature of scientific theories. Constructivism requires that theories be formulated in a way that permits constructive confirmation of their hypotheses; not all logically equivalent formulations are equivalent in this respect.

The obvious suggestion is that the auxiliary $\forall y(By \leftrightarrow Ty)$ requires some kind of independent justification from the evidence, and it is obvious from T2 that we can find all the instances we like of H2 and H3' without justifying that auxiliary at all. Under constructive confirmation, we cannot generate a legitimate constructive tree to confirm H1 in theory T2 simply by observing instances of H2 and H3', because to do so would violate hypothesis independence. We'd have to presume that *S*-ness was associated with *T*-ness in order to use the evidence to "justify" the claim that *S*-ness is associated with *T*-ness. To really justify H1, we need correlations between (for example) *B* and *T*, just as constructivism requires.

Edidin's example of a puzzling justification is again some limited evidence that confirmation is constructive. For suppose we can appeal to any hypothesis of any theory as an auxiliary without worrying about whether it is independently justified. Counter-intuitive cases like this would then be legitimate. We could use a consequence of T2, for example the consequence that $\forall x(Bx \leftrightarrow Tx)$, as an auxiliary to justify *H1* of T2. The Constructivist analysis gives the intuitively correct account of why we cannot make this move. To use $\forall x(Bx \leftrightarrow Tx)$ as an auxiliary, we need to confirm it independently of the target hypothesis, $\forall x(Sx \leftrightarrow Tx)$. But we cannot get this independent confirmation, because we are forced to presume that the target hypothesis is true.

Another example of violation of hypothesis independence comes from Christensen (1983, 474–475). Kepler's third law states that:

L3: The square of a planet's period (year), divided by the cube of its mean distance from the sun, is a constant for all planets.

Obviously, we ought not to be able to confirm this by observing only a single planet. Yet apparently we can do so. I'll use Mars as an example. Mars is, on average, about 226 million kilometers from the sun and takes 1.88 years to orbit, giving us a value of about 3×10^{-7} for the constant, or 3, using convenient units. Call this alleged constant 'k3(mars)'. I observe that Mars, in addition, obeys Kepler's first two laws. (Recall that the first law says that the orbit of a single planet is elliptical, and the second that it sweeps out equal areas in equal times.)

It is a consequence of Kepler's laws that:

KepAux: for any two planets, the first obeys both the first and second laws if and only if their k3 constant is equal.

And the rest is easy. From my observations of Mars alone, using KepAux, I infer that the k3 constant for Venus is also 3. Had I observed that Mars did not obey Kepler's first law, the same logic entails that the k3 constant of Venus is not 3. So I have confirmed that the k3 constant of Venus is 3, by observing Mars alone (Christensen 1983, 474–475).

Constructivism will object, naturally, that KepAux needs to be independently confirmed if it is to be legitimately used. How, then, could we gather evidence for it from what we know to have been observed? It will turn out that any way we invent for gathering evidence in its favor requires us to assume Kepler's third law (Fig. 7).

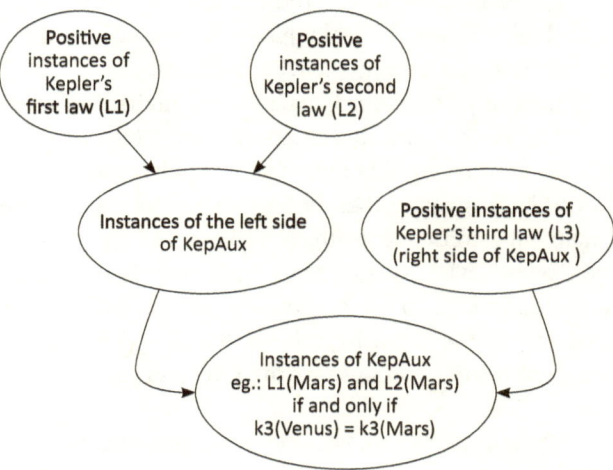

Figure 7: Christensen's KepAux Example.

KepAux is a biconditional. The example depends upon the idea that we can get true instances of the left-hand side, that is, that the planets obey Kepler's first two laws. So what we need in order to constructively confirm KepAux is positive instances of the right-hand side (negative instances would destroy Kepler's whole theory). But finding positive instances of the right-hand side is just confirming Kepler's third law. So the explanation for the failure of the example is just that in order to use it justifiably we must assume that the generalization it purportedly confirms is already true, and already confirmed sufficiently to justify accepting it as is true for all cases. That violates hypothesis independence. Once again, the case is ruled out if we require that there be some feasible prospect of getting to a constructive justification.

To summarize: You must give a constructive tree for the kepaux example if it is to be legitimate. To do that you require evidence for the auxiliary. To get evidence for the auxiliary, you must get evidence for the target hypothesis. Since it isn't possible to get evidence for the auxiliary hypothesis without assuming that we already possess evidence for the target, we do not have real confirmation. That is, it is impossible that auxiliary should be confirmed without violating hypothesis independence.

2.4 How to confirm auxiliaries independently using Bayesianism

We can break a constructive tree apart into the target justification, presented at the root node, and the sub-trees that justify each of the auxiliaries that we use there. A set of trees is called a forest. So we have an instance of relative confirmation, which is the target justification, and a forest, which together justifies each element of $\{A\}$.

Now suppose we are subjective Bayesians, so that our target instance of relative confirmation consists of a target hypothesis, H, the probability of which is raised by observing $\{E\}$ relative to $\{A\}$. Bayes' theorem again:

$$\Pr(H|\{E\}^\wedge\{A\}) = \frac{\Pr(\{E\}|H^\wedge\{A\})).\Pr(H|\{A\})}{\Pr(\{E\}|H^\wedge\{A\})).\Pr(H|\{A\}) + \Pr(\{E\}|\neg H^\wedge\{A\})).\Pr(\neg H|\{A\})}$$

The problem to be addressed is: What is the relationship between this instance of relative confirmation and the trees in the forest?

It is helpful to put the problem in terms of the subjective probabilities of an agent who has already made the observations that justify the auxiliaries and adopted the reasoning that they suggest. At the end of this process, the agent will have some probability distribution. What conditions does constructivism set upon this probability distribution if it is to count as a legitimate justification for the auxiliaries that appear in the above equation? What makes the justification of an auxiliary independent of a target hypothesis, or independent of which outcomes we actually observe?

There is an obvious way to adapt the intuitive suggestions about independent confirmation to Bayesianism. Hypothesis independence prohibits the justification for the auxiliaries from resulting in a belief in H. So the investigator cannot end up in this situation:

Constructive Bayesian hypothesis independence: Accepting the justifications for $\{A\}$ cannot result in $Pr(\neg H | \{A\}) = 0$.

This might be too strong, maybe there are acceptable examples of justification where there is a very slight confidence in $\neg H$ given $\{A\}$, but it is at least safe. The investigator cannot wind up with absolute confidence in H.

Evidence Independence follows similarly:

Constructive Bayesian evidence independence: Accepting the justifications for $\{A\}$ cannot result in $Pr(\{E\}|\{A\}) = 1$.

Again, this might be too strong, and might rule out some acceptable cases. But no acceptable case can be allowed to violate it. If $\{A\}$ is justified independently of the justification of H, then we cannot have the confirming observations be inevitable.

I have motivated hypothesis independence and observation independence in two ways. First, they intuitively capture the situation that real confirmation requires, and second, they explain why certain examples are not justifying, and how to make them justifying. But now, if we are Bayesians, we have a third reason for thinking that justifications for auxiliaries must possess this kind of independence. For, if either condition is violated, Bayesian justification becomes impossible.

Take hypothesis independence first. If the falsehood of the target hypothesis is guaranteed to be impossible before we begin, then the right summand in the denominator of Bayes' theorem, above, drops to zero, and our hypothesis has probability unity in advance of seeing which observation comes up. Similarly, if the confirming evidence must be inevitable if we are to be justified in depending on the auxiliaries at all, then the prior probability for H, should we "succeed in" confirming it with these auxiliaries, becomes identical to the posterior probability. Making the confirming observations cannot contribute to our confidence in H.

Bayes' theorem coincides in its requirements with the lesson of the examples, and with the intuitive motivation behind independent confirmation as a solution to real confirmation. This suggests that there is a real feature of scientific justification at work here. Just as constructivism insists, the justification for auxiliaries must leave some "room" for the event of observing confirming outcomes to contribute to the justification of that particular target hypothesis, and some "room" for potential refutation.

2.5 Constructive trees give the relevant background knowledge for a justification

I will next make three points about this way of defining the idea of an independent justification. First, these two conditions can be adapted to give a statement of how to accommodate the background knowledge, B, that a Bayesian justification relies upon when we look at particular examples. Second, it gives us a way quickly to test whether a proposed constructive tree passes muster as genuinely providing independent justifications for the auxiliaries. And third, it follows, curiously, that independent justification need *not* be a symmetric relation.

We normally see Bayes' theorem stated something like this:

$$\Pr(H|\{E\}^\wedge\{A\}) = \frac{\Pr(\{E\}|H^\wedge\{A\})).\Pr(H|\{A\})}{\Pr(\{E\}|H^\wedge\{A\})).\Pr(H|\{A\}) + \Pr(\{E\}|\neg H^\wedge\{A\})).\Pr(\neg H|\{A\})}$$

B is supposed to be the background knowledge. But it cannot be *all* the background knowledge. For one thing, as Glymour noted, the background knowledge sometimes includes the fact that E has already been observed (1980, 85–87). But if the evidence already has probability unity in the background, it cannot raise the probability of H. A new theory often gets support from past observations. Additionally, if H is already hugely probable, as for example in some vastly confirmed law, no new piece of evidence can be significant. H is now in B. $Pr(E|\neg H^\wedge B)$ will be zero (or undefined, if $\neg H^\wedge B$ is a contradiction) and $Pr(H|E^\wedge B)$ will be unity. So H gets no support from E. The trouble is, the new evidence might be some completely new way of testing H, and thus very significant. B somehow has to be the background knowledge that is *relevant*, but we aren't given a way to decide what's relevant within our general set of prior beliefs.

Constructivism offers help here. What should be in B, the constructivist says, are the hypotheses we cite in the constructive trees we use to justify $\{A\}$. It is these justifications that we need in order to make the target justification work. But we can often justify the As without citing the fact that we have already observed outcome $\{E\}$, as opposed to some $\neg E$. That is, we can justify the auxiliaries without using the old evidence that we cite in the target justification. Constructivism provides a way to estimate the probability that $\{E\}$ would be ascribed under the counter-factual supposition that it hasn't yet been observed. Howson and Urbach (1983, 404–406) propose this solution, without coming to any definite conclusion about how to constitute B.

To give an illustration of the constructivist solution: The precession of the orbit of Mercury can be evidence for General Relativity in spite of the fact that we observed that precession long before General Relativity was invented. We can confirm the value of the mass of the sun, distance between it and the sun, and the other auxiliaries, without depending upon our past observation of that precession. Then, in the target justification, we can compare the prior probability of $\{E\}$ given these justifications for $\{A\}$ and Newton, with the probability of $\{E\}$ given general relativity (so that H here is general relativity, and $\neg H$ is Newtonian mechanics). This is a way to estimate the probability an investigator would have for $\{E\}$ when he has some justification for $\{A\}$, but has not yet observed whether or not $\{E\}$, and is agnostic between the two theories of gravitation. Finally, show that the probability of H increases from this when our hypothetical investigator changes the probability of $\{E\}$ by actually making the observation.

There's a simple way to check that when a target justification is added in, it really does add something to the constructive trees justifying the auxiliaries. Think of a way that the target hypothesis could be false, even if the auxiliaries are all justified by some set of sub-trees. If that way in which a target hypothesis could be false is ruled out by the evidence in the target justification, then we have both hypothesis and evidence independence, and the resulting tree is legitimate.

For example, suppose we have some evidence that the lengths of objects a and b are identical. Suppose we got that evidence by rolling a cylinder along the objects and counting the same number of revolutions in each case. Is it an independent justification to measure the two with rulers? Or by timing the progress of inchworms along them? Well, look at the constructive tree we used to first justify measuring lengths by rolling cylinders. It could be that that tree checked the results of different cylinders against each other, and so leaves open the possibility that rotating an object through a circle increases or decreases its circumference, depending upon where the object is located. Then the rulers, or the inchworm, rules out this potential source of failure. So we do get independent confirmation that a is as long as b by these methods, even though normal humans might regard these as trivial repetitions. Once again, the background knowledge that is relevant is that which is in the constructive trees justifying the auxiliaries.

Now for a separate remark. I've been arguing that if both hypothesis independence and evidence independence hold in a constructive tree then we have a justification for some A in $\{A\}$ that is independent of the justification for H by$\{E\}$.

That sufficient condition for independent justification has an odd feature though. If *A* has a justification that is independent of the justification of *H*, it doesn't follow that *H* has one that is independent of *A*'s justification. Independent justification is not always symmetric. It is even possible that the target hypothesis entails at least one of the auxiliaries in a constructive tree confirming it, so long as the converse does not hold.

Constructive trees often "sneak up" on the target hypotheses. As we progress from leaf to root, we rule out more and more rival theories that would permit our target hypothesis to be false. At each stage, the hypothesis we confirm at that stage has evidence that both confirms it and rules out additional rivals to the target. Finally, at the target justification, the new evidence, and the auxiliaries, rule out another potential rival.[5] We will see examples of this when we ask how constructivism could ever get started in the next chapter.

I do not see this asymmetry as damaging to constructivism. On the contrary, ordinary examples in the sciences evince the same feature. We might have had justifications that receding light sources red-shift independently of the hypothesis that the universe is expanding, but no evidence for the expansion of the universe that is independent of the hypothesis that receding light sources red-shift. Perhaps, at one time, the only evidence for the expansion was the red-shift of the galaxies, but there was evidence for the red-shift from many sources. The same phenomenon will appear anywhere we have a lot of justification for hypothesis *X* but the only justification we possess for *Y* depends upon *X*. Then *X* is justified independently of *Y* but not conversely.

2.6 Cycles and why they are forbidden

Hypothesis independence prohibits any tree that contains the target hypothesis, *H*, as the background knowledge. This prohibition has a constructive Bayesian justification, as I have noted earlier. When we ask for the prior probability of *H* given the relevant background knowledge *B*, *B* must include all the hypotheses in the constructive trees we are using for *H* in this context. (The context matters because we might be asking what the probability of *H* would be if we do not know some things we do know, for example old evidence.) But if *B* includes *H* then (assuming consistency) $Pr(H/B) = 1$, and a successful observation cannot confirm *H* because it cannot raise its probability. As I have also noted, it follows

---

[5] I don't believe we ever rule out all the rivals, although often we rule out all that we can think up at the time.

that when *both H1* is confirmed *only* by trees that include *H2 and H2* confirmed *only* by trees that include *H1*, there is no constructive justification for either hypothesis.

It is important that the *only* available trees should result in the cycle. If either *H1* or *H2* is justified independently, we can use that fact to get some constructive justification for the other one. It is worth dwelling on this point.

Suppose first that we ask what the best case is for believing some specific hypothesis *H*. We, as it were, make *H* the figure, and make other hypotheses the ground. We invent all kinds of trees out of the outcomes of observations we know about, and as long as none of them violate either hypothesis or observation independence, they are all perfectly acceptable.

Now we drop the topic, and focus on another hypothesis *H'*. We ask for the best case for believing *that*. The answer, I'll suppose, is another vast collection of trees, so that *H'*, like *H*, is well justified.

Now suppose that some of the trees justifying *H* in our first question include *H'* and some of the trees in our second question contain *H*. Quine-Duhem thinks this situation is routine, and I agree.

Shouldn't constructivism immediately declare that these trees are cycles and so illegitimate? Aren't we then forced to depend only on the trees that confirm *H* independently of *H'* and vice versa? So won't it follow, as we consider more and more hypotheses in turn, that eventually more and more trees will get ruled out and constructivism will collapse?

No, because we have switched the context of each question. If we'd asked the question "What is the evidence for *H* that is independent of *H'*?" then we couldn't include, in the justification of *H*, trees that include *H'*. But we didn't ask for that. We asked, first, "what is the best case for *H*?" and then "what is the best case for *H'*?". These are not the same question, any more than the two commands "lift your left leg off the ground" and "lift your right leg off the ground" are really one command, so that nobody can stand on one leg.

The literature on scientific explanation is aware of the contrast class of the request for an explanation (van Fraassen 1980, 127, 224). "Explain why green plants emit oxygen in sunlight" might be asking why they emit oxygen, as opposed to carbon dioxide, or it might be asking why sunlight, as opposed to darkness, results in the emission of oxygen, or why green plants, as opposed to fungi, emit oxygen in sunlight. We get a different question, and different answers, depending upon what contrast we intend.[6] The same point holds for requests for justification. The

---

6 Why do birds fly south in winter? Because it's too far to walk. Why does Santa wear red suspenders? Because otherwise, his pants would fall down.

first question we asked was "why should I believe H, as opposed to disbelieving H?" and the second question was "why should I believe H' as opposed to disbelieving H'?" The latter question is not identical to "why should I believe H' as opposed to disbelieving both H' and H?" The contrast class is different.

"What reason is there for believing H that is independent of everything else I believe?" gets the answer "none whatsoever" from constructivism. But it *should* get that answer. It was a fantasy of foundationalism that there is some source for justification outside of science for the hypotheses that are inside of science. As Wilfred Sellars once observed, we can challenge any hypothesis we like because we can put it at risk of refutation, but we cannot put everything in jeopardy at once (1963 [1991], 170).

None of this touches the issue of cycles. For in those cases, the only justification for either of two hypotheses (or three or four . . .) requires the truth of one of the others. That might be tolerated as a temporary necessity, or as a promising line of research, or as highly suggestive, but skepticism is always legitimate. The skepticism may be slight, as we have good prospects for independent confirmation in future. But sometimes there are very strong reasons for thinking that no independent justification will ever happen. Under those circumstances, constructivism must say that skepticism about hypotheses in the cycle should be very strong. They have no constructive justification whatsoever, and miserable prospects of getting any.

This book argues that our possessing a constructive justification for H could be a necessary condition for our possessing a real justification of H. So far as we know, all real justifications might be constructive. While I do not have a knock-down argument to the effect that we lack any real justification when we lack a constructive justification, I will make the case that it is perfectly reasonable to think so. That is, it is perfectly reasonable to believe that only hypotheses with a constructive justification have any real justification.

Suppose that there is a cycle of justifications that does indeed provide us with a reason to believe some hypothesis. Constructivism cannot then be a necessary condition for real confirmation, since we have an example that violates hypothesis independence, and so cannot be constructive, but still provides a good reason to believe. It is therefore incumbent upon me to show that any examples of cycles of justification provide only a doubtful reason to believe.

2.7 Examples of cycles in the literature

Some of the examples of circular justifications in the literature are wildly counterintuitive. Peter Achinstein criticized Clark Glymour's bootstrap theory of confirmation by giving an example which was permitted by Glymour's original theory, but which permitted a spurious hypothesis – in this case concerning the quantity of God's attention – to be confirmed. But Achinstein's example requires a cycle of confirmation. Suppose, with Achinstein, that:

A = the total force acting on a particle,
B = the product of a particle's mass and acceleration
C = the quantity of God's attention focused on a particle

Achinstein's counterexample has two hypotheses:

(i)   A = C

(ii) B=C (Achinstein 1983, 170).

Here, we can confirm (i) using (ii) and conversely, but when we try to confirm either in its role as auxiliary, we are forced back on the other.

No other way is suggested to justify C. That was the point of the example, of course: to show that something very wild would be confirmed if Glymour's view were correct. So this gives us one of the reasons why we might suspect that some examples will continue to be unable to break out of the cycle in the future. There's a long history of attempts to discern evidence for God's nature and activities in the things we have observed. These have not fared well. So it is reasonable to continue to be pessimistic about future evidence. The same would be true about future evidence for phlogiston, or purposiveness in biology, or an aether that carries light.

Another example of a cycle is Christensen's example using Kepler's laws above. That was an attempt to confirm Kepler's third law, that the orbits of any two planets both obey a certain ratio, by looking at a single planet. It depended on the auxiliary that a single planet obeys Kepler's first and second laws if and only if two planets obey the third law (I called this auxiliary 'kepaux'). Here again, we got a cycle of confirmation. We cannot confirm kepaux without evidence for Kepler's third law. That is why the example looks so vacuous.

Cycles of justification can be extremely suggestive. Constructivism doesn't deny that it can be a striking coincidence that the evidence should be just such as to coincide with a cycle of justifications, and that that can look very promising as a program for future research.

But constructivism must say that no genuine justification has as yet been given by these cycles. According to constructivism, it must be reasonable to be skeptical. There must be something intuitively "wrong" with the cycle, even if some scientists are prepared to see it as convincing. The justified hypothesis must look insecure, unless there is some way to see it as justified constructively.

2.8 An Example: Newton's original idea for absolute velocity

What we are looking for is a circular justification that apparently provides real justification. There is a *kind* of example here that played a prominent role in the literature on this subject, and which turns out to be a cycle. That kind of example is of great philosophical interest, so that it's important to note it here. It's been very controversial, for reasons that have nothing to do with constructivism, and some philosophers have thought it to be a perfectly good kind of justification, while others have thought that it gives no justification whatsoever.

Consider Newton's first law: "Every body continues in its state of rest, or of uniform motion in a right line, unless it is compelled to change that state by forces impressed upon it" (Newton 1689 [1934], 13). Newton was well aware that bodies with no forces acting on them would be (apparently) observed to accelerate if the observer himself was accelerating – as we would now say, if the frame of reference is accelerated. He nonetheless maintained that in space itself, absolute space, bodies with no net forces acting on them obeyed the law.

For convenience of exposition, I will now suppose that the center of mass of the solar system is unaccelerated, and the fixed stars unaccelerated with respect to it. (The argument can be re-stated without this supposition, but is then more difficult to follow.)

There are hypotheses, for example that the fixed stars have no velocity, which Newton must say are either true or false, because of his famous remarks about absolute space and time:

> Absolute space, in its own nature, without relation to anything external, remains always similar and immovable. Relative space is some movable dimension or measure of the absolute spaces; which our senses determine by its position to bodies; and which is commonly taken for immovable space. . . (1689 [1934], 6).

So it must be either true or false that the fixed stars are stationary, but we cannot tell which. As Newton puts it ". . . the true and absolute motion of a body cannot be determined by the translation of it from those which only seem at rest; for the external bodies ought not only to appear at rest, but to be really at rest" (1689 [1934], 9).

Suppose, though, that we allow cycles of confirmation. Then we would have no difficulty in showing that, for example, Polaris was at absolute rest. For Sirius is at rest, and Polaris doesn't change position over time with respect to it. So Polaris is at rest too. We know that Sirius is at rest, because it is at rest with respect to the center of mass of the solar system, which we know to be at rest because it is at rest relative to Polaris.

The history of philosophy has given us various ways to spell out the intuitively worrying aspect of Newton's original proposal. Constructivism provides a new one. There are plenty of circular justifications that objects are at absolute rest, but no non-circular ones. Newton provided us with no way to escape these cycles of justification, and it is very difficult even to imagine how to successfully do so. Whatever unaccelerated thing we can observe, that thing could always be either at rest, or moving with an arbitrary constant velocity.

A later development, neo-Newtonian spacetime, provided a way to finesse the problem, in effect by translating between different inertial frames so that unaccelerated ones could all be physically equivalent (Sklar 1974, 202–206, 236–237; Friedman 1983, 87–92; Earman 1989, 33). It abandons Newton's original idea of absolute space, because no one of these frames is picked out as the "right" one – the unique one that is at rest. Neo-Newtonian spacetime depends upon conceptual developments that were unavailable in Newton's time. A constructivist can see it as superior precisely because it reformulated Newtonian mechanics in a way that avoided these inescapable cycles of confirmation. The only thing the outcomes of observation can differentiate is whether an observer's frame of reference is accelerated or not, not its position or velocity, and the only distinction neo-Newtonian spacetime makes is between accelerated and unaccelerated frames of reference, not between frames at different positions or velocities.

One very natural objection to constructivism is to argue that cycles are never inescapable, because it is always possible that some future development will allow us to confirm the auxiliary independently. Several authors, for example, have proposed that it is possible that the Michaelson–Morley experiments had detected the aether. They have argued that we would be able to detect velocity with respect to Newton's absolute space by identifying the rest frame of the aether with the rest frame of absolute space (Sklar 1974, 196; van Fraassen 1980, 49–50; Friedman 1983, 115).

So, the argument goes, we might have suggested early on that the center of mass of the solar-system was at rest in absolute space, and used that hypothesis to confirm absolute velocities. That would result in a cycle, since we could only confirm that the center of mass is at rest by depending upon experiments which presumed that it was. But the cycle is not inescapable. Later, when we confirm that it really *is* at rest by showing that it is stationary in the aether, we break out of the cycle. The same argument applies to all cycles, so no cycle is inescapable.[7]

---

[7] Sklar discusses a number of puzzles about this argument, which I will pass over because they concern the specifics of the example and not the general challenge that no cycle is inescapable (Sklar 1974, 196–198).

We do not need to survey any possible future science to judge a cycle to be inescapable. To say that a cycle is inescapable is to say that, according to the hypotheses we know to have been confirmed now, we cannot see how to escape it. The worry is particularly trenchant when we know of a very well-established set of hypotheses that prevent any experiment from ever independently establishing the hypotheses. I mean such things as the Heisenberg uncertainty relations, or the conservation of energy, and linear and angular momentum. When such things prevent independent confirmation of hypotheses, the cycle ought to look very suspicious.

In Newton's case, two features are particularly worrying. First, as I have mentioned, whatever unaccelerated thing we pick that we can observe, it might always be either at rest or moving with constant velocity. Second, the way that Newton uses absolute space, to solve the puzzle of ruling out what we would now call accelerated frames of reference, itself apparently rules out the possibility of ever breaking out of the cycle. For suppose different positions in absolute space at different times affect the physical behavior of bodies in such a way as to allow us to discriminate different velocities. This contradicts the indistinguishable behavior that his first law attributes to bodies that are at rest, as compared to moving at constant velocity. If we were to be able to break out of the cycle by observing motions that allow us to distinguish different absolute positions at different times for unaccelerated bodies, then Newton's first law would be false. Constructivism must say that there's something *methodologically* wrong with a theory which maintains some hypothesis, when *that very theory* entails that "justifications" for it must be circular.

That isn't a completely decisive argument. There might be something other than a difference in motions that allows us to draw the distinction between rest and uniform motion; perhaps different bodies would be different colors at different velocities, for example. The argument is nonetheless extremely disconcerting. For the very reason that the colors are not motions, it's hard to see how they are connected to locations at different times. It's difficult to see how to confirm non-circularly that the different colors are associated with different speeds in absolute space. We can apparently add or subtract an arbitrary velocity northwards without affecting anything.

In historical situations in which we discover an inescapable cycle, the scientists of the day ought to have looked on the situation with dissatisfaction. They must have been worried about how to get some independent justification for the hypotheses. One can "confirm" Newton's absolute positions, intervals and velocities, but only if one supposes other hypotheses stating that inertial frames of reference are absolutely stationary, or assigns them a fixed velocity. Once one has done that, one can use the observed behavior of bodies to "confirm" absolute position. That is a cycle of course, and scientists and philosophers have worried about it ever since Newton made his proposal. Even Newton reads as though he has some reservations (1934 [1686], 12).

One can avoid a cycle by reformulating a theory as well as by breaking out of it. We never did discover a way of constructively confirming that an inertial frame was at rest in absolute space. But we found a way to treat each inertial frame as equivalent, and to formulate a theory that didn't need hypotheses about absolute space. Constructivism can explain the advantage of this theory over Newton's original formulation – it avoids hypotheses which could not be constructively confirmed. Constructivism doesn't make claims about meaning that require us to say that Newton's absolute space was meaningless. But as a hypothesis about real confirmation, it does say that Newton's absolute space, as he formulated that notion, gave rise to worries about whether it could move to real confirmation even with the passage of time. Other theories where constructive confirmation looks inconsistent with strongly justified hypotheses can be in the same boat.

3 How to Confirm Hypotheses about Unobservables

There are two mutually reinforcing arguments for the Quine-Duhem hypothesis. One, Duhem's, is that any judgment that observation has had some outcome depends upon the truth of additional hypotheses – observation is theory-laden (Duhem 1954 (1982)), 180–190). We may choose to abandon one of these additional hypotheses if we do not like the implications of the observation for some target hypothesis. The other, present in Duhem but emphasized more by Hempel, is the focus of this chapter. It concerns the use of hypotheses concerning observable entities to confirm or refute hypotheses concerning unobservables (Hempel 1966).

In order for us to believe that the antics of observable things are manifestations of unobservable things, we must depend upon some hypothesis about how the unobservables relate to the observables. Only by depending upon such a hypothesis can we confirm any hypothesis about the unobservables. If the outcomes of observation come out as predicted, then hypotheses about the unobservables are confirmed, and hence their existence. But we have to begin with this initial hypothesis linking unobservable entities to the outcomes of observation. So if we dislike the consequences of observations for some target hypothesis, we can abandon or modify this initial hypothesis instead of abandoning or modifying the target.

This initial foothold could not be justified by the outcomes of observation. For it was (allegedly) a precondition for justifying any hypothesis about unobservables. Since it was itself a hypothesis about unobservables, it could not be justified until it had been justified.

Notoriously, writers such as Carnap and Reichenbach tried to say that the initial hypothesis was a matter of meaning, or stipulation, or convention. The initial hypothesis was a coordinative definition (Reichenbach 1958, 14–19), or analytic hypothesis (Carnap 1935 [1996], 53; Carnap and Gardner 1966, 257–274), or correspondence rule (Nagel 1979, 97–105).

In "Two Dogmas of Empiricism" Quine argued that no empiricist could privilege a hypothesis in this way (1953 [1980]). None of our linguistic behavior noncircularly shows where we draw the analytic/synthetic distinction, nor even that we draw it. Most seriously, for philosophy of science, science *does not in fact* set aside hypotheses as irrefutable by observation. Since that is so, we will always have a choice about which hypothesis to abandon when the observations do not behave as we expect; we could abandon the initial hypothesis about how unobservables manifest themselves, or we can abandon the hypotheses about unobservables that we use it to confirm. The Quine-Duhem hypothesis is secure.

This chapter argues that it is not secure. Quine's picture of science established that there was a multiplicity of means by which hypotheses may be confirmed, and it severed empirical justification from issues of meaning and ontological parsimony. Once we have agreed to these, we can still keep the idea that we begin by establishing that an unobservable manifests itself in some way before we go on to confirm other hypotheses about it.

How does such a process get started? This chapter proposes a way. The very features Quine proposed – a multiplicity of means of confirmation and an indifference to parsimony – themselves provide a means to confirm some initial hypothesis about how an unobservable manifests itself. Induction by instances can confirm a "redundantly expressed" hypothesis about unobservables without using any analytic definitions. An example is: "If a and b are laid alongside, then a is as long as b when and only when the ends coincide". This initial hypothesis can be confirmed by itself, without requiring additional hypotheses concerning unobservables, by observing consistent results of its truth over time. The logical positivists overlooked this possibility, presumably because they were so concerned with the most economical means of expressing hypotheses.

Such hypotheses are by no means uncommon. They are richly scattered across every empirical discipline. Making use of them initially, we may confirm successive, and richer, hypotheses about unobservables, as this chapter goes on to show.

Clark Glymour proposed a very similar idea, and this chapter begins by reviewing it. The rest of the chapter adapts a variant of his idea to the Bayesian theory of relative confirmation. The first example is a banal kind of "scientific theory". It shows how the workings of a ruler to detect otherwise unobservable differences in length can be justified. In more complex theories, where the behavior of one unobservable depends upon another, we can also find by observation a condition in which one unobservable is alleged by the larger theory to be absent, and use it to get an initial foothold to justify subsidiary hypotheses concerning another. Then, by varying the conditions, we can expand the justification to the other unobservable entity. The theory of the balance – a special case of the law of the lever – illustrates this.

This proposal is only supposed to get the justification of unobservables started using a restricted set of data. It is an original acquisition problem. It does not show that we can eventually get to the position of having any strong confidence in them. Other developments later in the book, particularly fault-tracing and the chapters on observation, show how this initial confidence could be amplified.

3.1 Glymour's idea

Clark Glymour (1980, 111) gave one solution to the problem of confirming hypotheses about unobservables from outcomes of observation. Suppose we wish to confirm one version of the gas laws. When in equilibrium under stable conditions the pressure, volume and temperature of a gas are always related by the equation:

$$PV = kT$$

where we do not know the value of *k*, a constant. What do we do to find it? Well, we observe one set of values for pressure, *P*, volume, *V* and temperature, *T* for a sample of gas. Then we calculate the constant from those. Call these values *P1*, *V1*, and *T1*:

1. $k = P1V1/T1$.

Now, intuitively, this doesn't confirm the gas law. Whatever values pressure, volume and temperature are observed to possess, we will get a value for *k*, and so the gas law is not put at risk. But if we observe *two* values for each of the quantities, then we can put 1. at risk. The two calculations for *k* clearly might conflict. It is thus *some* evidence for the gas law that varying (for example) temperature and pressure will result in a value for the volume that gives the same constant as before.

Some will say, at this juncture, that we only have evidence for the constancy of some kind of ratio:

2. $P1V1/T1 = P2V2/T2$.

There's no denying that 2. will account for the data so far. Indeed, it accounts for future experiments with the same sample of gas.

But consider some other ways to express 1.:

3. $PV = nRT$

where *n* is the number of moles of gas, and *R* is a constant, the *gas constant*. Or:

4. $PV = NRT/N_A$

where *N* is the number of particles and N_A is Avogadro's constant.

Different samples of the same gas give different results. But different samples of the same gas *with the same mass* give the same result not only for *PV/T*, but for the actual values of *P* and *V* at different *T*'s. That is, under these varying circumstances (comparing two samples of a single gas each having the same mass) the constant of proportionality is invariant too. Different gases, though, still have different constants of proportionality. But now, again, we can extend the class of circumstances under which the outcome is predictable if we use the number of moles of gas to make the prediction. As each of these different variables allows us to cope with more and more cases, so also, at least intuitively, the idea of hypotheses about unobservables being true becomes more and more attractive.

Jan Zytkow (1986) presented two other examples of this kind of process in action. First, the conservation of linear momentum and second, drawing upon Herbert Simon (1970), Ohm's law. The second is particularly interesting for my purposes because of the care with which Simon traced the sources of background knowledge. Ohm worked in a context in which previous experiments had demonstrated (he thought) that he had a reliable means of measuring current (i) and resistance (R). Current was measured through a force, that between a coil and a magnet, mounted on a spring and using Hooke's law. Ohm varied resistance by the length of a piece of wire inserted in the circuit (Simon 1970, 17). Ohm then demonstrated the law that bears his name:

$$\exists v \exists b \forall i \forall R (i = v/[b + R]).$$

Here, v is what we'd now call the voltage, or electro-motive-force, and b is a constant, the internal resistance of the battery. Ohm could not observe these at the time, even indirectly, except through the very relation of Ohm's law. So here we have a hypothesis with two unobservables, b and v, which depends upon background knowledge in a clearly articulated way. If we can just confirm Hooke's law and some analysis of forces, plus some analysis of the length of the wire, we can get Ohm's law from these.

Glymour's original idea was to confirm a hypothesis by using that hypothesis itself as an auxiliary. Then, of course, if something goes wrong, we cannot blame an auxiliary in preference to the target hypothesis because the two hypotheses are identical. By the same token, we are not required to assume without evidence that the unobservables must show up in the observations in some specified way. The way an unobservable manifests itself in observations and a hypothesis about that unobservable could be the same hypothesis, and so the outcomes of observations can justify this single hypothesis. That is an extremely ingenious idea about the way in which we might have originally justified hypotheses about unobservables.

It is difficult to use that idea directly within a Bayesian system of relative confirmation. We cannot, for example, justify a hypothesis when that hypothesis itself is in the background knowledge, because it's prior probability with respect to that background is unity. And we cannot confirm it when it is not in the background, because if we do not include some statement about unobservables in the background we have no way of linking the outcomes of observation to any hypothesis about unobservables. Of course, we could try including some different hypothesis about the link between observations and the unobservables in the background, but then we'd just be stuck with the Quine-Duhem problem again. It's difficult to adapt the idea to constructivism too, because it looks as if we will be depending upon the truth of the hypothesis to justify it, and a cycle of justification threatens.

Still, I think there's a way of adapting the idea using another of its features. Hypotheses concerning unobservables have many different manifestations in the world of observables. If we can get started even slightly by using one way in which unobservables link observed initial conditions to predicted outcomes, we could amplify our initial confidence that the unobservable entity exists by showing that it links other observations together too.

3.2 Adapting Glymour's idea

The question at issue is: how could anyone possibly get started in confirming hypotheses about unobservable entities given only observations? The example I will propose concerns length. Imagine someone who has not been trained in using rulers, micrometers, surveyor's wheels, dividers, or any other way to precisely measure length. Suppose this person is skeptical about whether there is any more precise physical quantity than that which is evident to the unaided eye (call it 'eyeball-length'). 'Is as short as', this person thinks, might be something inherently vague like 'has as short a temper as'. What sequence of experiments could ever convince this skeptic? Nobody, after all, can use eyeball-length to *see* that a ruler is more accurate.

The essential idea is that unobservables get confirmed when they support an open-ended sequence of correlations between observable outcomes. The hypothesis that unobservables exist does not only predict merely a single observable regularity. Nor does it only predict the observations we have actually made. Nor does it only predict the regularities we would expect to continue if we abjured unobservables and proceeded by enumerative induction from observable instances which we have observed. Hypotheses concerning unobservables go beyond this and say that we will observe new regularities. This was the moral of the demise of operationalism, of course (Hempel 1965, 123–134).

Persuading our skeptic follows this pattern. The initial experiments predict only hypotheses concerning the "unobservable" concept of length measured by rulers. Several of these weakly confirmed hypotheses predict new regularities, which then get confirmed. In what sense 'new' regularities? In the sense that our skeptic about more-exact-length has no reason to expect those outcomes given only the experiments we have already displayed (these yield nodes lower down in the tree). Our skeptic could find these successes surprising, although we will find them banal. Nobody is forced to see the outcomes that confirm the existence of unobservables as inevitable, given only the generalizations from the data with which we began. This sense of novelty is extremely impoverished, but it's good enough to get started.

Stathis Psillos (1999, 173) mentions what I think is the same idea. The hypotheses concerning unobservables, when contrasted with a refusal to take them as justified by observable regularities, predict something novel, or surprising. They predict correlations among observable correlations. If we observe these, we have a success that was not duplicated by simply taking observable outcomes to be bare regularities with no further structure.

3.3 A toy example

I draw upon the work of van Fraassen (1980, 1989) to defend the idea that objects, properties, and processes are sometimes observable, sometimes unobservable, and that in saying this we are treating the human body as a kind of measuring instrument (1980, 13–19). For convenience, suppose we are provided with two large samples of conveniently manipulated objects with straight edges – the birthday cards of a brother and sister, for example. So we can observe these, name them and reidentify them, and observe the coincidence of end-points when they are laid alongside each other.

Begin with just the boy's cards. Consider this hypothesis:

5. $\forall t$ if one end of two boy-birthday-cards, a and b, are juxtaposed at t, then (length(a)=length(b) if and only if a, b coincide at the other end at t)

I take *length(x)* to be something unobservable, but the other properties are observable. Can our skeptic confirm this by repeatedly juxtaposing cards *a* and *b*?

Yes, because as far as the skeptic is concerned, juxtaposing the ends of straight edges might give different results each time – the skeptic is not familiar

3.3 A toy example

with the operations of rulers, and the eyeball-length of some object is only approximate. So there are three rival possibilities:
a) Coincidence of end-points is not repeatable.
b) Coincidence of end-points is repeatable, but not due to this *length* thing.
c) Coincidence of end-points is repeatable, due to this *length* thing.

The evidence speaks against a). It doesn't discriminate between b) and c). Being a good Bayesian, then, our skeptic slowly raises the probability of both b) and c) as the probability of a) declines with increasing evidence (assuming non-extreme priors).

(To illustrate: suppose for simplicity that a) means that the coincidences will *never* repeat, and the skeptic assigns prior probabilities of 1/3 to each alternative. Then prior to a test of repetition of coincidences, $Pr(c)=1/3$. By Bayes:

$$Pr(c|E) = \frac{Pr(E|c).Pr(c)}{Pr(E|c).Pr(c) + Pr(E|b).Pr(b) + Pr(E|a).Pr(a)}$$

$$= \frac{1 \times 1/3}{1 \times 1/3 + 1 \times 1/3 + 0 \times 1/3}$$

$$= 1/2.$$

After repeating the experiment, the new probability of $Pr(c)$ becomes the old probability of $Pr(c|E)$. So the probability of *c)* rises from 1/3 to 1/2, and *c)* is confirmed. This will work as long as we have non-extreme prior probabilities, and observing the regularity decreases the probability of hypothesis *a)*.)

Now vary the experiment slightly. Use the birthday card named *b* as a "designated ruler for boy-birthday-cards".[8] So two cards are the same length, as measured by *b*, if they both reach as far along *b* when laid alongside it. Gathering instances, we have evidence confirming:

6. ∀t ∀x ∀y if one end of each of two boy-birthday-cards, y and x, are juxtaposed with ruler b at t, then

    (length(y)=length(x) if and only if the ends of x and y, reach the same point on b at t).

Once again, I am using the skepticism of the skeptic to argue for the significance of observations which are, to us, completely insignificant. *We* know that if card *a* always reaches the same point along *b* when *b* is taken as a ruler, then

---

[8] Preferably, of course, pick the card with the longest side.

every card will do the same. But the skeptic doesn't know this. Maybe card *a* is a special one, which when repeatedly laid alongside *b*, reaches the same point on *b*, but other cards don't. So long as skeptic ascribes a non-extreme probability to this, hypothesis 6 rises in probability as a result of the experiments. For at least one of the rivals to 6 gets ruled out, namely that *a* is an exceptional card.

In this way, *b* will categorize the cards into classes that are as long as each other. That is, if there is such a thing a length. For all our skeptic knows, these results *might* only hold of the boy-birthday-cards. Card *b* might be a special card too, so that we couldn't get this result from other birthday cards. Of course, if this *length* thing is real, the results will be more general, but we do not know that yet.

Now do the same set of experiments on the girl-birthday-cards. This time, pick an arbitrary card called *g* and use it as the "designated ruler for girl-birthday-cards". Gather evidence favoring:

7. $\forall t \, \forall y \forall x$ if one end of two girl-birthday-cards, x and y, are juxtaposed with g at t, then

   (length(x)=length(y) if and only if y, z coincide with a single point on g at t).

So *g* will again partition the set of girl-birthday-cards into sets with the same, and different, lengths.

So there is some increase in probability for both 6 and 7, although not much. We have, in Bayesian terms, confirmed them.

But once again, given the evidence we have looked at so far, a great deal of skepticism is still available. The skeptic might wonder whether the relations of "same length" that we established in the two categories of birthday cards are really identical. Of course, if 6 and 7 are true, then, if cards *c* and *d* are the same length according to ruler *b*, they ought also to be the same length according to ruler *g*. We do not know this *a priori*. The "same length as" relation might be like the "taxed as much as" relation, which varies from country to country, but is systematic within a single country.

The hypothesis:

8. length(c) = length(d)

is a consequence of the observations and 6 (that is, cards *c* and *d* were birthday cards addressed to the boy, and measured using *b*). We can confirm it independently by using hypothesis 7, that is, by measuring it using ruler *g*. When we succeed, that rules out a rival to the idea that objects have some unobservable kind of length that is more exact than we can detect by eye. So we can confirm 8 using the skeleton (Fig. 8).

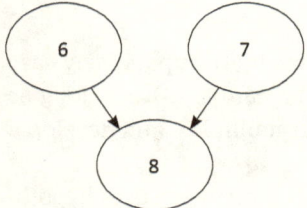

Figure 8: Identity of Length.

Hypotheses 6 and 7 were confirmed directly from observable outcomes without additional auxiliaries: boy-cards and girl-cards repeatedly reached the same points on each designated ruler. Then 8 is a new prediction (though trivially so, from our point of view), which is confirmed when we observe it. This is enough to get justification of hypotheses concerning unobservables started. It depends only on outcomes we can observe. So constructive Bayesianism can at least get to the first rung of the ladder.

Let's take a step back. Look at the logical form of 6. and 7. We have confirmed simply by induction a hypothesis expressed by a sentence of the form Carnap called a bilateral reduction sentence (1953, 54):

6. $\forall x \forall y \ \forall t \ (O1xyt \rightarrow (Uxy \leftrightarrow O2xyt))$

Here, $O1$ and $O2$ are predicates picking out some property we can observe of observable objects, while U concerns some property humans cannot observe that observable objects possess. We can confirm this just by observing correlations between $O1$ and $O2$.

Then we confirm another, of the same logical form, in which the unobservable is responsible for the correlation between two different observables, $O3$ and $O4$:

7. $\forall x \forall y \forall t (O3xyt \rightarrow (Uxy \leftrightarrow O4xyt))$

Once again, this is confirmed by induction, by observing only that $O3$ and $O4$ are correlated.

6. and 7. obviously state something beyond the observations that confirm them, but every case of induction does that. They together entail, though, something that a more modest generalization, omitting the unobservable, does not. To confirm that an object, a, exemplifies the unobservable property, establish that a

exemplifies *O1* and *O2*, and predict that it will exemplify *O4* when tested. This confirms that:

8. Uab.

If this is an acceptable strategy for getting started, then it is comparatively easy to see how to adapt it to confirming that one side of a card is twice as long as some other edge. From there, we could confirm other multiples, and develop a scale of length. This will turn out to be important.

I hope it goes without saying that the fact that constructivism uses hypotheses that Carnap would associate with his bilateral reduction sentences does not entail that it endorses other views that Carnap (or anyone else) held about their role. There is certainly no suggestion that such sentences need state the meaning of unobservables, or that they constitute the "empirical content" of a theory, or that they provide the resources for any version of reductionism.

3.4 Various objections and replies

3.4.1 Constructivism is contrived

This objection says: "This case is all wildly artificial. No real historical case proceeds like this. The constructivist is saying that *if* evidence and speculation proceeded in this order, *then* we'd have evidence for unobservables. But in most cases, this is a ludicrous parody of history."

The fact that historically justifications do not proceed in the order in which constructivism presents them is beside the point. The burden constructivism bears is to show that science would be possible if people did require real confirmation to be constructive. So it must show that important kinds of hypotheses can be justified constructively, not that they were historically justified that way. The story above shows how someone could at least begin with agnosticism or doubt about unobservables, and end with a stronger belief as a result of the outcomes of observation.

We constantly begin anew in evaluating and inventing justifications. We constantly reason from different sub-sets of the observations we know to have occurred. That is what we do when we are trying to figure out what has gone wrong with an experiment, or what could account for an unusual and troubling observation. It happens, too, when we are trying to present a topic in a systematic way for teaching purposes. It's also what happens when some new phenomenon is investigated, such as X-rays or the early stages of what became radio transmission. I believe it is also what we are doing when we try to think of independent evidence for a hypothesis, or criticize an experimental design. I will try to make this plausible as the book progresses.

3.4.2 The objection from van Fraassen's work

"What the constructivist has done is simply link pairs of observable regularities with others. Even though doing so favors some hypotheses about how unobservables behave over others, it does nothing against someone who begins with agnosticism about unobservables. Suppose someone assigns a prior probability of 1/2 to the hypothesis that the world will look as if it contains unobservables, but in fact there aren't any. At the end of this process, that hypothesis still has probability 1/2."

I grant the whole argument. This isn't supposed to show that constructivism is better justified than van Fraassen's constructive empiricism. It's supposed to show how unobservables in constructivism could gain some initial justification. I will address my case against constructive empiricism in a later chapter.

3.4.3 Constructivism is too permissive

"I can attach unobservables to any set of regularities that I know about. Every spring, the flowers bloom, and every time I push shopping cart, a wheel squeaks. So I can invent some bogus unobservable that is confirmed by these observable regularities in exactly the same way."

This challenge assumes that any unobservable that can get started is bound to end up as an intuitively compelling end-point of justification. There's an obvious reason why an unobservable that was introduced for the sake of shopping-carts and botany looks silly, namely, that we have established no independently confirmed subsequent hypotheses linking the two, whereas we have for the link between mosquito bites and malaria.

In replying in this way, I'm granting that an unobservable that is responsible for squeaky wheels and blooming flowers *does* get a start by observable correlations. I do grant this apparently absurd view. As Hume remarked, for all we know in advance of observation, anything may cause anything. By the same token, I would argue, any observable correlation might turn out to be linked by some unobservable to any other. It is when we consider what *else* we have observed of the world, and the reasoning that has enjoyed independent success with the evidence subsequently, that we winnow out hypotheses like witchcraft and astrology. The question to be answered in this chapter was an original acquisition problem. It asks how we could possibly get started. It doesn't address the issue of how to evaluate competing hypotheses each with some evidence in its favor.

3.4.4 The example takes for granted many hypotheses, so it isn't really a real confirmation at all

"There are still lots of hypotheses on which rulers depend which this alleged demonstration hasn't established. For example, it assumes that there isn't some temperature gradient that shrinks and expands objects so that they have no fixed length at all over time (Poincaré 1902, 65; Reichenbach 1958, 37). If this temperate gradient is there, the evidence doesn't really show that the rulers measure length. So the demonstration ought to begin by showing that it isn't there if it's going to work. It obviously doesn't."

How can we justify these assumptions, upon which the whole theory of ruler-length depends? Reichenbach says that we cannot, and that they are just arbitrary decisions (1958, 19). Carnap appeals to practical convenience (Carnap & Gardener 1966, 92–95). In both cases no evidence is presented, and as Quine argued, empiricism fails.

Part of the answer is that the presence of such a temperature field can be detected later on, when more evidence is available. There's another answer that is also illuminating though.

The objection makes confirmation by observation too much like mathematical proof. Justification from observation is always defeasible – there is always the potential that the hypothesis that is confirmed turns out to be false, in spite of the outcomes of observation. (It can also turn out that the judgment that that particular outcome occurred was a mistake.) It is in the nature of any applicable theory of confirmation to allow for this. We do not have to rule out every possible defeating rival theory in order to confirm a hypothesis in science.

A hypothesis, together with the background given by a constructive tree, is confirmed if we observe one outcome, and refuted if we observe some other outcome. This is something which we know from inspecting the tree, independently of contemplating which outcome actually happens. Some theory of relative confirmation – Bayesianism in this case – describes this situation. It also describes what happens when we contemplate the additional information that one outcome, and not another, actually happens. There is no contemporary theory of relative confirmation – certainly not Bayesianism – that is prevented from working by the gloomy thought that its verdict might be reversed as we observe more and more. Nor does that gloomy thought prevent Bayesianism from describing the initial situation prior to our contemplating which outcome happens.

3.4.5 These initial steps are a very poor justification

"The initial acquisition is simply too pathetic to count as confirmation at all. We are told that this is a story about how some longer tale of justification begins merely from the outcomes of observation. But the initial "justification" here is contemptible. No real person concludes things on such slender justification, that leaves so much open to future research. Constructivism cannot link hypotheses to finite sets of observable outcomes, because the story it tells about the first stages is simply too poor."

To which I reply that the opponent cannot have it both ways. You cannot, that is, object to foundationalism while requiring an initial acquisition of unobservables which is at all strong. Foundationalism required strong justifications from the initial stages of empirical justification. But the candidates for these strong initial steps – sense data or things like them – simply do not exist. We are forced to conclude that stronger justifications emerge from weaker ones. They emerge from the accumulation and spontaneous agreement of different independent justifications from the evidence. So if we grant the failure of foundationalism, we cannot go on to complain that the initial steps in a justification provide a rather shabby account of themselves. That is simply inevitable.

3.4.6 Justification only begins holistically

"We only get any sort of justification – not just intuitively good ones – when this vast mountain of data and justification has been amassed."

I have several answers to this.

First, it doesn't seem to me to be true. When John Snow hypothesized that unobservables in the water supply cause cholera because the distribution of cholera centered on a water-pump, and then removed the handle from the pump and observed the predicted vanishing of the disease, then that seems to me to be pretty good evidence for his hypothesis. It is certainly evidence in the case of the observable, but merely unobserved. I conjecture that the mysterious vanishing of food from the counters and the access of the dog to the kitchen are correlated due to his unobserved proclivity to steal it. I might observe him doing so. Or I might exclude him from the kitchen for a few days. The latter might well be convincing when I fail at the former.

Second, the objection proposes a rival to constructivism (maybe – see the next reply), but does nothing to establish that rival. We'd like to know exactly how independent justifications work, if they aren't built up as the constructivist suggests. To find out which position is better, we ought really to look at whether each view is justified by the observations we make of the science we possess. It isn't inevitable that they would both succeed.

Third, the objection threatens to degrade into a dispute about what is enough for something to deserve the word 'justification'. We might discover that the correlation we conjecture between pairs of observables doesn't hold. So the outcomes might refute our hypothesis. I think that the possibility of subsequent refutation like that is enough for something to deserve the term 'justification' when it succeeds rather than fails. But if you disagree, then I'll rephrase my case as follows: O.K. I'll agree to your assertion – we get justifications only by amassing huge amounts of data with many different patterns of reasoning. If you think that parts of this are not justifications, then I reply that you can use some other name for them. They are still something which, when accumulated, leads to an emerging conviction on the part of humans. We still need to know how to get these 'smustifications'.

One can rephrase the constructivist case this way, but one ought to bear something in mind when doing so. While it is merely a verbal dispute whether the word 'justification' should be reserved only for something that considers a large body of available data, or whether small bodies of data can provide smaller justifications, something else is not verbal. It is not a verbal dispute whether inescapable cycles of justification genuinely confirm. What one ends up believing will vary depending one what one says about this, in a way that doesn't vanish when we change which words we use for the same things. It is the refusal to allow inescapable cycles that really sets constructivism off from alternatives, not the fact that it starts small and works up.

3.5 The scales of justice

We just saw the original constructive justification for using rulers to measure length. Next we will see a quite different constructive justification for measuring length in the scales of justice. The two justifications are largely independent; quite different observations would have refuted most of the constructive tree of either without affecting the other. In particular, nothing in their justifications requires them to agree about which hypotheses concerning the length of specific objects are confirmed and refuted. In the next chapter, we will see that the two methods agree about some hypotheses concerning length, but disagree about others, and we will look at the way that new experiments and observations can track down where the error lies.

This example makes four other points in addition. The balance, or scales of justice, concerns two unobservable quantities, length and weight. It is a special case of the law of the lever, which says that a lever is in equilibrium if and only if a force, applied to one end of a lever, multiplied by its distance to the fulcrum, is equal to the opposite force at the other end, multiplied by its distance to the fulcrum. The scales of justice balance under these conditions. If one can measure forces, one can determine distances, and conversely. But apparently, one cannot use the theory to determine both together. This chapter shows that this natural conclusion is false. We can exploit the symmetries of the theory to justify constructively the measurement of each of the quantities in succession. So the second point this chapter makes is that constructive confirmation is sometimes possible even for a theory with more than one interdependent unobservable quantity.

The third point the example makes is that the same methods that work for justifying rulers, which are only doubtfully part of science, will work also for more realistic scientific theories. The law of the lever cannot be excluded from the practice of science. It is frequently the first example of scientific reasoning to which schoolchildren are exposed in the science classroom. Until the development of the mass spectrometer in the early 20th century, refined versions of the scales of justice were essential pieces of apparatus in Chemistry (and in all other natural sciences too). There was no more accurate way to determine the forces due to mass. It was the first theory Mach considered in his *The Science of Mechanics* (1893, 11). Mach went on to describe experiments justifying the whole of classical mechanics from this starting-point. Those familiar with his book will, I think, be able to see how constructivism could follow a similar path.

And fourth, lives have – literally – hung in the balance since ancient times. Murders have been committed over the readings of this instrument, and with its aid. It would be worth investigating whether and how these ancients could have been justified.

3.6 Apparatus

Consider the apparatus in figure nine. This version includes pans that can be moved back and forth along the armature, to vary the distance between the intersection of the pan and the armature and the fulcrum. I will assume that the position of the intersection can be marked on the armature, and I'll refer to this point as 'the intersection' of the pan for short. I also assume that we are provided with a variety of bodies each of which can be freely moved between the two pans, and collections of which can be placed on either pan. The apparatus balances when the pointer coincides with the benchmark and can be easily disturbed to tilt in either direction by a brush of the hand, and subsequently returns (eventually) to its original position. It is not part of the argument of this chapter to say where this apparatus comes from. The argument of this book shows a way in which such balances can be designed and improved (Fig. 9).

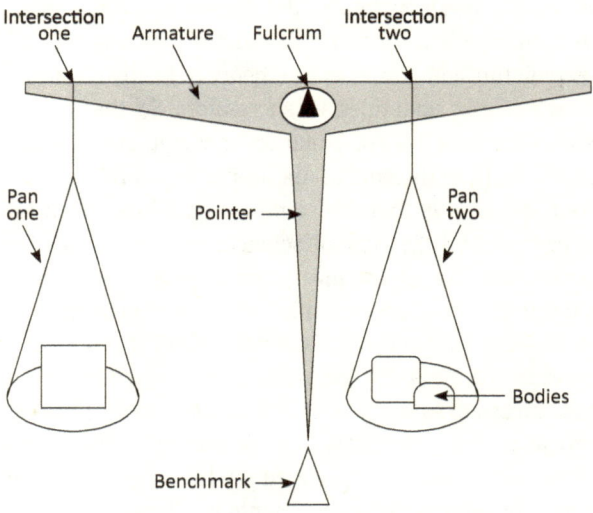

Figure 9: The Scales of justice.

3.7 The theory

A theory of this apparatus has also been known since antiquity. It is that the scale is in balance if and only if the products of the total weight of the bodies on each pan and the distance between its intersection and the fulcrum are equal. In symbols:

$$T: \forall t \forall x \forall y \forall p_l \forall p_r (B\,x\,y\,p_l\,p_r\,t \leftrightarrow d(p_l) \cdot w(x) = d(p_r) \cdot w(y))$$

In English, this says "At any time t, balance obtains between two bodies, x and y, (where x is on the left side of the balance and y on the right) at two points on the armature, p_l and p_r if and only if:

The product of the distance between x's position and the fulcrum with the weight of x is equal to the product of y's distance with the weight of y."

This requires some explanation. The variables 't' and 'p' range over times and positions of the pans on the armature respectively. It will sometimes be useful to use these letters as names as well as variables. The five-place predicate letter 'B' indicates whether the scale balances with objects on (respectively) the left and the right pan, at intersection p_l at left and p_r at right, at time t. Please note that the side upon which the body is placed is indicated by the position of its name after the predicate 'B', and the same goes for the distances to the pans on the left and right sides along the armature. The functors 'd' and 'w' refer to distance and weight respectively, where distance is taken always to be measured to the fulcrum.

For the apparatus to give good results, the armature must move freely, and the bodies whose weights are being measured must be of an appropriate size. The armature should be level when no bodies are in the pans. These do not exhaust the conditions you need to avoid error. The next chapter shows how to detect new sources of error, and ways to correct them, by fault-tracing.

3.8 List of observables

Guided (once again) by van Fraassen's distinction between observable and unobservable, I will take the following to be decidable as a matter of observation:

We can observe, at any time, whether balance obtains or not.

We can re-identify the bodies, the pans and other gross features of the apparatus.

We can observe which of two marks on a single arm of the armature is closer to the fulcrum.

We can observe whether or not the intersection of a pan coincides with a mark on the armature.

I take the following *not* to be decidable by observation:

Whether two arbitrary bodies have the same, or different, weights.

Whether a length between an intersection (or mark) on one arm and the fulcrum is greater than, less than, or equal to the distance between an intersection (or mark) on the other arm and the fulcrum.

If, after the theory has been confirmed, one wants to describe an observation of balance at equal distance of two bodies as being an observation of their equal weight, then I have no objection to offer (see, for example, Maxwell 1962; Shapere 1982; Kosso 1989a). If one wants to do this *before* the theory has been confirmed, though, my objection is that this 'observation' takes for granted the truth of hypotheses for which we as yet possess no evidence.

3.9 Confirming that two intersections are equidistant from the fulcrum

It is difficult to see how the balance can differentiate hypotheses concerning weights from hypotheses concerning distances. Confirming a hypothesis about weight, but not length, or vice versa, seems to require some assumption fixing the value of the other quantity. Both quantities are unobservable, so we cannot empirically determine either value directly. Hence, we appear to be forced to depend upon some assumption that is not established by observation. (Hasok Chang presents a very similar argument in the course of arguing that we must assume principles that are justified "neither by logic, nor by experience" (2004, 91) in order to measure any quantity that is not directly observable (2004, 59, 89–92; see also Chang 2001a).)

In order to make the exposition easier, I will make a somewhat artificial assumption. Suppose the bodies are a,b,c, etc. Then I'll assume that some pairs of these bodies are equal to each other in weight. I'll also assume that the armature comes with points marked on it, *p1, p2, p3* etc. and that at least some of these are equidistant from the fulcrum. Then the problem is to identify which pairs of points are equidistant, and which pairs of bodies are equal in weight, using the apparatus and various hypotheses. One can do without this assumption by depending on dumb luck and fiddly experimentation. It's completely unilluminating to do so, however.

As I mentioned, we can exploit the symmetries of the theory to devise a test for when the distances from the fulcrum to two intersections are equal. If we can once justify that that identity obtains and is stable over time, then we can use the identity of distances to compare the weights of different bodies. So the first task is to get evidence that two points, which I will call p and p', are the same distance from the fulcrum. In what follows, points without the prime marker are taken to be on the left-hand side of the balance, and the prime indicates the right-hand side.

Two intersections, p and p' are equidistant from the fulcrum when at least one object is *exchangeable* at those intersections. Two objects are said to be *exchangeable* at intersections *p1* and *p2* if and only if they both balance each other, and continue to balance when the one on the left-hand pan is exchanged for the one on the right and *vice versa*. Now clearly, if there are two objects that are exchangeable, then *p1* and *p2* are the same distance from the fulcrum, according to T, and this is a test we can observe.

Now confirm:

1. $\forall t_e \forall t_l \forall x \forall p_m \forall p_n (Baxp_n p_m{'} t_e \rightarrow$
   $(d(p_n) = d(p_m{'}) \leftrightarrow Bxap_n p_m{'} t_l))$

(The variables t_e and t_l are meant to suggest an earlier and a later time).

In English: use object *a* as a 'test object'. This first experiment has two stages. First, pick two points on the armature, p_n and p_m'. Second, holding these points fixed, put *a* on the left pan first and another object, *o*, on the right. If balance obtains, trade *a* and *o*. If balance still obtains, we have a temporal instance of 1. Confirm that 1 is so by repeating the experiment.

For most pairs of points, no object is exchangeable with *a*. According to hypothesis 1, then, these pairs of points are at different distances from the fulcrum. For some pairs of points a few objects are repeatedly exchangeable. The hypothesis entails that these points are equidistant from the fulcrum. (With our background knowledge, we know this consequence is true, but no experiment shows that yet.)

Now justify hypothesis 2:

2. $\forall t_e \forall t_1 \forall x \forall p_m \forall p_n (Bxbp_n p_m' t_e \rightarrow$
$$(d(p_n) = d(p_m') \leftrightarrow Bbxp_n p_m' t_1))$$

We are using a different test object, *b*, and putting it initially on the right side of the balance rather than the left. We find we can confirm 2, so that *b* again labels some pairs of points as equidistant, and other pairs as not.

Nothing so far shows that two different test bodies should result in the same pairs of points being equidistant. That is, it's consistent with all we have observed so far that *b* should show that *p* and *p'* are equidistant, but that *a* should show (when *p* and *p'* are used for hypothesis 1) that they aren't. But suppose *p* and *p'* are equidistant according to *b*. By 1, then, we predict that *a* will balance on exchange with some objects at those points, if the points really are equidistant. We find this is true. So we have confirmed:

3. $d(p) = d(p')$

3.10 Confirming two bodies have identical weight

Next confirm, by the time stability of balance, that

4. $\forall t \forall x (w(d) = w(x) \leftrightarrow Bdxpp't)$

This simply involves showing that a new test body, *d*, consistently balances with anything it once balances with (and the same for imbalance).

Now consider a body, *e*, that balances with *d* when *e* is on the right pan. Search around to find a body, *f*, that will balance[9] *e* when *e* is on the left-hand pan. By time-stability, confirm:

5. $\forall t(w(f) = w(e) \leftrightarrow \text{Befpp}'t)$

Now confirm:

6. $w(d) = w(e)$

by putting *d* in the left pan and *f* in the right, and observing balance. No observation, or hypothesis, so far shows that the 'same weight' relation is transitive. But hypothesis 6, if added to 4 and 5, entails that *d* will balance *f* when *d* is on the left and *f* is on the right. That is what we observe, so 6 is confirmed using 4 and 5 as auxiliaries.

Thus far, we have shown that bodies *a* and *b* are exchangeable between the pans. Let *Dxy* (for "x direction-balances y") be the property of *x* balancing *y* when *x* is on the left pan and *y* is on the right pan, at *p* and *p'*, but with no commitment to the reverse being true. We have observed *Dde*, *Def*, and *Ddf*. For all we know so far, with the exception of bodies *a* and *b*, it could be quite false that *Dxy* entails *Dyx* when we have equidistant points of intersection for the pans and the armature. But if we add this hypothesis:

7. $\forall t \forall x \forall y (\text{Bxypp}'t \leftrightarrow w(x) = w(y))$

then it follows that we will observe *Ded*, *Dfe* and *Dfd*. We do observe these things, which means that 7 has been confirmed independently of its auxiliaries.

Hypothesis 7 is something like a working theorem in mathematics. It actually gets put to work in confirming all kinds of other hypotheses. Other theorems may be useful as intermediaries in getting to theorems like this, but rarely show up again in future developments. As a preliminary to the next step, we can use 7 to establish the identity and difference of the weights of multiple bodies.

3.11 Establishing a scale: Integer multiples and fractions of weight

For the next step, I am going to depend upon the theoretical nature of our observations. Introduce the functor '∘' to mean 'the body formed when two bodies are

---

9 The argument will also work if *f* produces imbalance.

3.11 Establishing a scale: Integer multiples and fractions of weight — 79

mixed or otherwise closely juxtaposed'. I will also use this notation when I want to talk about two bodies on the same pan, so that 'Ba∘b c p_l p_r t' means that we get balance when both *a* and *b* are on the left pan and *c* is on the right-hand pan.

The functor '∘' is not part of theory *T*. It follows that we need to add a hypothesis to *T* that says how '∘' behaves. I add this hypothesis:

$$\forall x \forall y w(x \circ y) = w(x) + w(y).$$

Just adding the hypothesis doesn't confirm it. We can understand it and use it to make predictions without knowing whether or not it is true. The objective is to confirm this additional hypothesis.

To do so, confirm:

8. $\forall x \forall z \forall w \forall t \forall t' \ ((Bxzpp't \& Bywpp't) \rightarrow$
$(w(x \circ y) = w(x) + w(y) \leftrightarrow Bx \circ y z \circ w \ pp't'))$

That is, get two pairs of bodies that are the same weight. Use stability over time to show that when one body from each of the pairs is placed on one pan, their weight is additive if and only if balance obtains when and only when the other bodies are placed together on the other pan.

I now want to confirm:

9. $\forall x \forall y w(x \circ y) = w(x) + w(y)$

8, along with the observations, entails that pairs of bodies have additive weight, but it says nothing about triples. To establish 9, then, find three pairs of bodies that have the same weight as each other, and use 8 twice to predict that the triples will balance. Hypothesis 9 is entailed by the discovery that they do balance, and would be refuted by the discovery that they did not.

This isn't a very thorough confirmation of 9. Once you had tentatively established 9 this way, you would naturally want to gather additional evidence substantiating it. You would want, for example, to confirm it by the following method.

Put two bodies on the left pan. Find a cup that is of negligible weight. Put the cup on the other pan and fill it with water until it balances the pair. Then separate the body of water into two bodies by pouring part of it onto one pan until it balances just one of the two bodies. (Thus body of water *J* consists of bodies of water *l* and *k*, so that *j* = *l* ∘ *k*.) Hypothesis 9 predicts that the remaining water balances the other body, and if it didn't, we would have a counterinstance to 9. From hypothesis 9, it is easy to see how to establish fractions and multiples of weight, and to develop a scale of weight with standard bodies.

3.12 Confirming a scale of length

Because we will now be dealing with multiple points of intersection which are not equidistant, relabel point p as p_1 and p' as p_2.

Identify three bodies r, s and t all of equal weight. By hypotheses seven and nine, $w(r \circ s) = 2w(t)$. Now confirm:

10. $\forall t \forall x \forall y (Bxyp_1p_3t \leftrightarrow (w(x) = 2w(y) \leftrightarrow 2d(p_1) = d(p_3)))$

as follows.

Carefully mark the positions p_1 and p_2 so that they can be found again later in the experiments. Put t on the right-hand pan and r and s on the left. Now move the right-hand pan across the armature until it balances the left. Mark the position. This is p_3. Now confirm the left-to-right portion of the embedded biconditional in 10:

$$\forall t \forall x \forall y (Bxyp_1p_3t \rightarrow (w(x) = 2w(y) \rightarrow 2d(p_1) = d(p_3)))$$

by using the time variable and the reidentification of the same point p_3, which is (according to the theory) twice as far along the armature as p_2. (That is, simply repeat the experiment at different times and discover that the same point is selected.) Use a variety of different triples of bodies all having the same weight, as established by hypothesis 7, in order to get the universal generalizations over bodies. (The attentive reader will note that, given 7 and 9, it is possible to establish that a body has double the weight of another, so that it is not always necessary to use triples of bodies.)

Now confirm the embedded converse:

$$\forall t \forall x \forall y (Bxyp_1p_3t \rightarrow (2d(p_1) = d(p_3) \rightarrow w(x) = 2w(y))).$$

Find two new bodies, m and n, which balance when the pans are at $p1$ and $p3$. Then use a scale of weight and hypothesis 7 to confirm that $w(m) = 2w(n)$. (Other experiments will also work; you could confirm contrapositives, for example.)

Now we have a position p_3 on the right side of the armature that the theory says is double the distance from the fulcrum that p_1 and p_2 were. The next step is to follow exactly the same procedure in mirror image to establish a fourth position, p_4 on the left side. Confirm:

11. $\forall t \forall x \forall y (Bxyp_4p_2t \leftrightarrow (w(x) = 2w(y) \leftrightarrow 2d(p_2) = d(p_4)))$

as for hypothesis 10 *mutatis mutandis*.

Now confirm T:

T. $\forall t \forall x \forall y \forall z \forall w (Bxyzwt \leftrightarrow (w(x) \cdot d(z) = w(y) \cdot d(w))$.

Move the pans so that they are at positions p_3 and p_4. If T is true, these should be the same distance from the fulcrum. Now take two objects that have been established by hypothesis 7 to have the same weight and put them in the pans. By hypothesis 9, 10, plus the observation that balance obtains with equal weights at p_3 and p_4, an instance of T follows. If we observe that balance does not obtain, we have a counterinstance of T. So T is confirmed.

I think the reader can easily see how the example just given can be generalized for other instances of theory T. For example, we can double or half the distances at which bodies balance, or we can use four bodies and confirm versions of 9 and 10 for multiples of three. The interested reader can devise scales of weight and length.

3.13 Overview of the balance

Note that the observation of weight need not have anything to do with the way bodies feel to us when we are handling them in a gravitational field. We could have a justified belief in many hypotheses about weight without our ever having felt the forces that the bodies exert upon our hands. Scientific concepts need not begin with the subjective phenomena that they intuitively cause or describe.

This sequence of experiments need only increase our confidence in the hypotheses involved in the theory of the balance very slightly. All the same, it performs as advertised. It confirms the theory progressively, where each new hypothesis is confirmed by the outcome of observations and independently of the preceding body of confirmed hypotheses. In doing so, it uses the observed outcomes of different experiments to confirm the background knowledge it needs to eventually confirm hypotheses about unobservable entities.

This entire process of justification is a constructive tree with the skeleton shown in Fig. 10.

82 — 3 How to Confirm Hypotheses about Unobservables

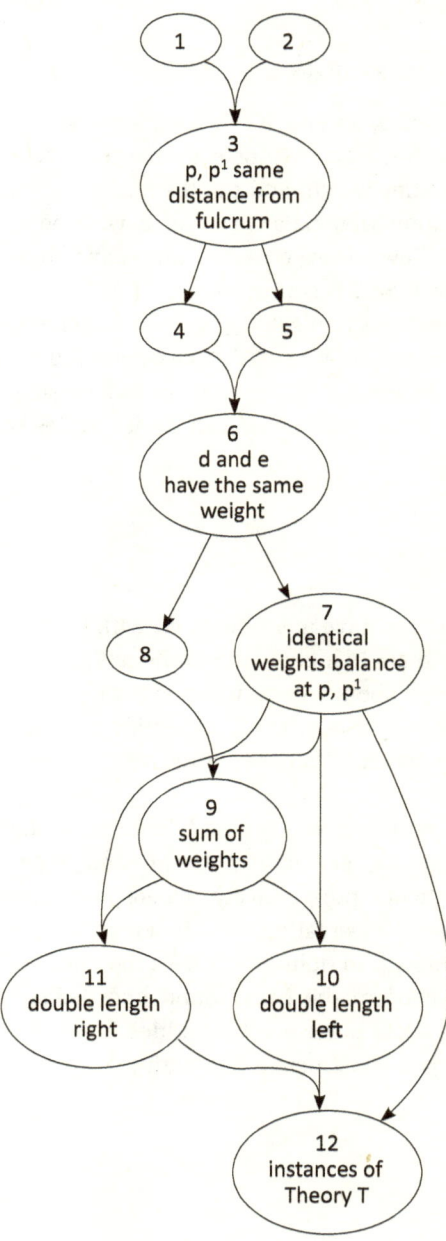

Figure 10: Skeleton of the Constructive Tree for the Law of the Lever.

4 Fault Tracing

This chapter shows by example that science contains, as a normal and familiar part of its functioning, a method that allows us to use the outcomes of observation to discover whether or not a given hypothesis is responsible for a failed prediction. If the Quine-Duhem hypothesis is to have any relevance to the practice or philosophy of science, it must deny this. So the present chapter gives a counterexample to Quine-Duhem, and works through that counterexample in practice.

4.1 A nasty surprise

We now have two observably distinct methods for measuring length, the ruler and the balance. We have amassed a variety of observations using these devices. We can use these observations to confirm constructively the hypotheses required to think of them as independent methods for measuring length.

With that starting point, the obvious thing to do is to see whether or not a scale of length established by one of these methods matches one produced by the other. And as soon as we do this, we are brought to a sudden halt by the fact that it doesn't.

The ruler and the balance always agree about which objects are the same length. When the measurements of the balance say that two pans are equidistant from the fulcrum, the ruler agrees that they are. But when it comes to *multiples* of length, they disagree. Suppose the balance says that object a is twice as long as object b. The ruler will then say that object a is slightly less than twice as long as object b. Conversely, of course, if the ruler says that c is twice as long as d, the balance says that c is more than twice as long.

4.2 What might have gone wrong?

What's striking is that the two methods actually agree on many hypotheses concerning length. A hypothesis confirmed using hypotheses of one theory is independently confirmed using hypotheses of the other. One way to proceed is to begin with the tentative thought that the hypotheses upon which there is agreement could well be correct. We have now two independent constructive trees confirming each of them, after all.

Following this hopeful thought, we suspect that we do at least know when two lengths are identical, because the theories independently confirm these identities. We can develop a scale of weight using only these "safe" hypotheses. That is, we could follow the constructive tree for the balance up to hypothesis 9, where we could determine sums of weights and develop a scale of weight.

The hopeful thought that we know when two lengths are identical is supported by at least two additional arguments. First, while our eyes are quite inaccurate ways to measure length, they are not useless. Particularly for fairly short distances, the lengths the balance says are multiples just do not look correct, whereas the lengths the ruler says are multiples do. Our eyes suggest that the ruler has got lengths correct and the balance has got them wrong, although both methods look correct when they say that two distances are equal.

Secondly, the justification for hypothesis 10 in the case of the scale was highly suspect. (This was the first hypothesis, using the balance that did not depend on the hypothesis that two lengths are identical.) We found three bodies of equal weight, and put two on one pan. Then we put the third on the other pan and moved it along the armature until it balanced. We checked this in various ways, for example by showing that 3 other equal weights would balance in the same way at the same points on the armature. We then just concluded that the distance must be double. But we only used hypotheses about identity and difference in *weight* to do this. We never measured the length in a way that didn't presume that it would behave according to the proportions of weights. Clearly, we were just "taking the theory's word for it" that this was *double* the distance. It might have been triple, or 1.6 fold, or any other factor.

So, following this reasoning, tentatively conclude that we can rely on the constructive tree for the balance for identity of weight, and sums of weight for different bodies. So we still have a scale of weight, and we are tentatively supposing that rulers are accurate about a scale of length too. What does that suggest about which hypothesis is at fault?

The errors follow a detectable pattern from this perspective. Table 1 is a (slightly artificial) table, using an arbitrary scale of weight.

In the left-hand column, we have the weights of each of the three bodies that are established by the balance to be equal. We put one of these on the left-hand pan, at intersection p_1, and the other two on the right-hand pan. Then we find the point of balance by moving the right-hand pan to p_3. The right-hand column shows the result of using the ruler to compare the lengths at which the scale balances with two of these weights on one side and one on the other.

Now the ratio of lengths ought to be 2 in each case. That is, when the weights of the three equal bodies, with two on one side of the balance and one on the other, are all 1 unit, the ratio of lengths at balance should be 2:1, but instead is 3:2. When each body weighs 2 units, the ratio should be 2:1 – that is

Table 1: Ratios of lengths for different weights on the balance.

Weight of each of the three bodies	Ratio of lengths at balance points
1	3/2
2	5/3
3	7/4
4	9/5
9	19/10

4:2 – but turns out to be 5:3. When each body weighs 3 units, the ratio should be 6:3, but is instead 7:4, and so on.

In short, when the balance says that:

$$\text{length}(a) = 2\,\text{length}(b),$$

the ruler (and our estimate by eye) say that instead there is a constant length, k, such that:

$$k + \text{length}(a) = k + (2\,\text{length}(b)).$$

This is highly suggestive. It suggests that, on each of the two pans, there is some kind of "secret weight". (The secret weight is of one unit in the above table.)

If we compensate for the secret weight, by making the bodies 1 unit lighter on each side, we ought the get the factor of 2 for the lengths. This "secret weight hypothesis" gets support from the fact that when there are no bodies on the pan, the balance-points are equidistant. The secret weight is identical on either side, so it contributes the same additional length to both sides. That agrees with the observation that the ruler agrees with the balance in the hypothesis that equal weights would balance at equal distances.

And it's not so hard to figure out what the secret weight is; it's the weight of the pan and the side of the armature.

4.3 Independently confirming that that is what went wrong

This is probably the most important part of the theory of fault-tracing. I have not found the point discussed explicitly elsewhere in the literature on confirmation, nor in philosophy more generally.

When we have constructive trees, refuting data does not bear on every hypothesis that is formerly confirmed "as a single, undifferentiated whole", as Glymour once put it (1980, 3). In this case, the two constructive trees, plus the comparison between the two methods, show that the balance and the ruler agree with respect to some hypotheses, and disagree with respect to others.

We have found something – the weight of the armature and the pans – such that *if* it had been what went wrong, it would have saved our favored theory. It would "make the results come out right". But we do not just declare victory at this point. We seek some way to test whether or not that really *is* what went wrong, or whether something else might have been at fault.

We try to confirm, independently of other hypotheses and trees that were involved in generating the original problem, that the suspect hypothesis really is guilty. If constructivism is correct, we seek new evidence (or new ways of reasoning from evidence we already have) that constructively refutes the hypothesis which we suspect is the culprit. And we seek to do this independently of other hypotheses that might be at fault, to guard against their falsehood indicating that an innocent hypothesis is guilty. We also commonly try to find independent indications that the hypotheses we think are innocent really are.

What does it mean to say that the suspect result was confirmed to be correct *independently of hypotheses and trees that gave rise to the original problem*? In this case, and many others, some hypotheses, of each theory, are reinforced by an independent justification from the other theory. The hypothesis that p and p' are equidistant from the fulcrum, for example, is reinforced by measuring the distance with a ruler. Others are contested between the two theories in conflict. Refutations, like justifications, are relevant only to some hypotheses of our theories, and throw only some into doubt. So the italicized phrase means at least that, to get evidence that the suspect hypothesis really is at fault, we require some refutation of it that depends only on hypotheses for which the two theories agree.

It's desirable in addition that we get a constructive tree that is independent of even the well-confirmed hypotheses involved in the background of the original sequence of experiments. When we have got at least minimal independent evidence for the location of the fault, so that our beliefs and observations agree, I say that we have *restored consonance between our beliefs and the observations*.

I think it is evident from this example and others I will give in the next chapter that we do in fact check up on the correctness of some diagnosis of the faulty hypothesis. I emphasize here that this is an observation we make about how we do science. Constructivism can give an account of why we accept the independent justifications we do, and why we reject at least some others. That account depends upon localizing the chains of justification for our hypotheses and bringing them to an end. If constructivism is wrong, it's very difficult to see how to do the same thing.

So how could we check up on the fault in this example? We have suggested that the error lies in the assumption that the *only* weights on the pans are those of the bodies that rest on them. But in fact, the weight of the pan itself, and the armature, also affects matters.

The best approach would be to establish the reliability of some new method for measuring weight – a spring balance, for example. We could then weight the armature and pans, and see if the result coincided with the prediction in the amended version of theory T that compensates for the error.

One isn't forced to develop a new way to measure weight like this, though. It is also possible to use the independently confirmed and "safe" hypotheses to restore consonance. Using algebra, the reliability of the ruler, and the fact that we have a well-confirmed scale of weight using the balance, it is possible to calculate what the weights of the pan and armature should be if this is what is really at fault. Then one can weigh the pan and armature using a different balance with the pans at equidistant points. This depends on some of the hypotheses in T, but only those that got a lot of independent confirmation the first time around. It isn't perfect, but it'll do.

We used the suspected location for the error to make a prediction, which we then tested using hypotheses for which we had additional evidence. This prediction came out correct. That is some evidence for the location of the error, although no guarantee.[10] I repeat the central point: We didn't just stop with our initial guess at where the error lay. We used it to make predictions about tests that are independent of the suspect hypothesis and new to the experiments we have got so far. A kind of novelty is involved. The outcomes of these new tests might either convict or acquit our suspect hypothesis given the observations we possess and the reasoning we regarded as secure. So their coming out correctly supports the location of the flaw in the theory. We have restored consonance between the observations and our beliefs.

4.4 In some cases, the observations prevent a choice of blame

Sometimes, it is just false that there's a choice about which hypothesis is to blame when a correction is made to a theory. Every alternative view about what went wrong runs into problems with the observations.

---

[10] We would like some evidence that other locations for the fault do not succeed here. We can get this by reasoning from the hypotheses we have used in the trees so far. It's very difficult to alter them in a way that also succeeds.

I gave three examples in the Introduction to this book. The error responsible for the Ultraviolet Catastrophe includes the hypothesis that the energy and frequency of light are independent. The error responsible for the null results of the Michaelson-Morley experiments includes the invariance of length and time with change of reference frame. And Lord Kelvin erroneously assumed that atoms in the earth release no energy over time when he criticized Darwin's theory.

Our current evidence strongly supports the hypothesis that frequency and energy are not independent, that both time dilation and length contraction are required for the round-trip experiments with light, and that some atoms in the earth release energy over time as they decay radioactively. It insists, moreover, that we have excellent evidence that the fault lay in denying *these* hypotheses, and did not lie in some other hypothesis. There is better evidence that Kelvin was wrong about the continuous release of energy by the earth than there is for any alternative.

I do not claim that the justifications from the data are infallible. What I claim is that the data that we actually know about strongly confirm that the error lies where we believe it does. They confirm this particular error and refute the idea that the problem is due to any other hypothesis. Evidence can locate what hypothesis is at fault, and it can do so in a way that denies us a choice.

It is all very well to *say* that we can hold onto any hypothesis come what may, without cost to the data, but it is very difficult actually to show that this is true of real scientific examples. It's particularly germane that it's the *data* that alternatives conflict with when one tries to locate the fault somewhere else. It is not pragmatic concerns such as the simplicity of the theory, or the number of phenomena it can explain. This suggests that it is the outcomes of observations that we use to select whether to believe a hypothesis, not these pragmatic concerns.

The opponent to constructivism makes an existential claim. He claims that there are ways to distribute doubt among hypotheses that will undermine the justification of any hypothesis without conflicting with the data. When we ask our opponents actually to produce these redistributions for examples like these, we find they cannot do so. The existential claim that blame can be redistributed is justified indirectly, by alleging that justifications in the science we possess are relative. That allegation cannot survive an articulated alternative that gives these examples their natural reading and says we have used the evidence to convict the guilty hypotheses and acquit the innocent. That is why these examples of inescapable correction to hypotheses of a theory appears to me to be so important. Since the indirect argument from the relativity of confirmation is no longer available, the supporters of the Quine-Duhem hypothesis owe us a more detailed account of how to evade these examples. I am very doubtful that one can be produced.

Appendix: What about Mercury?

The orbit of Mercury failed to fit the predictions of Newtonian mechanics for two hundred years (Lakatos 1978; Lakatos and Musgrave 1970; Kuhn 1962 [1996], 39, 81). Constructivism holds that science contains, as a normal part of its ordinary functioning, a method for discovering whether or not a given hypothesis is responsible for the failure of a prediction. But if that is so, why couldn't the Newtonians discover the flaw in Newtonian mechanics? This looks like a failed prediction for constructivism, because there was an enormous effort to solve this problem over a very long period of time, so that is looks as if the guilty hypothesis ought to have been identified. Moreover, the example is a paradigm for the Quine-Duhem hypothesis. It appears to fit exactly the strategy of blaming auxiliaries to protect the favored hypotheses.

Lakatos, for example, used the example of Mercury to defend Quine-Duhem:

> . . .the most admired scientific theories simply fail to forbid any observable state of affairs. (1978, 16)

He wrote:

> . . . some scientific theories are normally interpreted as containing a ceteris paribus clause: in such cases it is always a specific theory *together* with this clause which may be refuted. But such a refutation is inconsequential for the *specific* theory under test because by replacing the ceteris paribus clause by a different one the *specific* theory can always be retained whatever the tests say. (1978, 18, emphasis in original)

The specific theories that cannot be refuted include all our most admired ones. So no evidence can refute any of our most admired theories. We can always blame less-favored hypotheses in exactly the way that Quine and Duhem argued.

That is the argument in bald outline, but it misses a lot of its persuasive power. A much longer quotation gives a clearer sense of the way in which Lakatos pressed his case. He considered a physicist who uses Newton's theory, N, initial conditions, and background knowledge to calculate the orbit of a small planet p:

> . . .the planet deviates from the calculated path. Does our Newtonian Physicist consider that the deviation was forbidden by Newton's theory and therefore that, once established, it refutes N? No. He suggests that there must be a hitherto unknown planet p' which perturbs the path of p. . . . then asks an experimental astronomer to test his hypothesis. The experimental astronomer applies for a grant . . . In three years' time the new telescope is ready. Were the unknown planet to be discovered, it would be hailed as a new victory for Newtonian science. But it is not. Does our scientists abandon Newton's theory. . .? No. He suggests a cloud of cosmic dust hides the planet from us . . . and asks for a research grant to send up a satellite . . . Were the satellite's instruments . . . to record the existence of the conjectural cloud, the result would be hailed as an outstanding victory for

> Newtonian science. But the cloud is not found. Does our scientist abandon Newton's theory . . .?
> No. He suggests that there is some magnetic field . . . (1978, 17, ellipses added)

Lakatos continues in this vein for some time. He concludes that "either yet another ingenious auxiliary is hypothesis is produced or . . . the whole story is buried in the dusty volumes of periodicals and the story is never mentioned again" (1978, 17, ellipsis in original).

As the reader can no doubt anticipate, my objection to Lakatos is that he ignores restoring consonance as a feature of fault-tracing. When the example is viewed in the light of this feature, it does not show what Lakatos thinks. We require evidence that a blamed hypothesis is really at fault. We require this evidence to justify its target independent of the hypothesis we are trying to save.

What the example of Mercury really shows is that we cannot always find this evidence. Sometimes we simply cannot confirm which hypothesis is at fault. Constructivism succeeds because it can show that there are some cases in which we can use independent justifications to show that a hypothesis is innocent or guilty. But it is no part of the view that we can always do this. Sometimes the evidence, or our ingenuity, is uncooperative, even for centuries.

The constructive answer to the example

There are features of this example which support the constructivist answer here. First, if the Quine-Duhem were correct, one would expect that the problem of Mercury would not have appeared to be a problem for so long. And second, if it does persist as a problem, and Quine-Duhem were wrong, one would eventually expect it to provoke symptoms of disquiet, as there is some evidence that it did.

Lakatos' argument claims that Newtonian mechanics forbids nothing about the data. If Lakatos is to defend this, then he must say that *eventually* we would succeed in accounting for the anomaly without altering the core set of Newton's views. He has to say this, because otherwise Newtonian mechanics *would* constantly be in conflict with the data. To be sure, we might not abandon it, but we'd be forced to admit that it had a problem with the evidence that it couldn't (so far) solve. We must, that is, eventually be able to find a way to avoid the refutation of the theory by the anomaly.

But nothing in the story that Lakatos tells us explicitly guarantees such an outcome. Indeed, the evidence he points out seems to indicate the reverse, that sometimes we *cannot* get the evidence to cooperate, and Newtonian mechanics *does* forbid certain observations.

There is an additional problem as well. For, under Quine-Duhem theories of justification, why bother even to *try* to get this subsidiary evidence for some diagnosis of the problem? A non-constructivist says that there is nothing wrong with two hypotheses mutually confirming each other, so that each requires the other as a presumption when the evidence confirms it. It might be nice, the non-constructivist says, if there were independent confirmation for one or the other, but this isn't required in order for the two to gain justification from the evidence. (This assumes, rather generously, that the non-constructivist can make sense of the idea of independent justifications.)

It appears, then, that we could solve the problem of the anomaly in the orbit of Mercury as follows. First, use the evidence of the anomalous orbit of Mercury, plus Newton's law of gravitation, to confirm that there is a planet close enough to the orbit of Mercury that perturbs that orbit. And second, use the existence of that planet, *without having actually observed it*, plus the orbit of Mercury, to confirm Newton's law of gravitation. This doesn't appear to offer any support for either, though. Just as the constructivist alleges, it is a cycle of justification using the planet and Newton's law of gravitation.

Perhaps the example is unfair to the non-constructivist. After all, that position doesn't say that *every* example of mutual confirmation provides support. Perhaps the mutual support is too tight, or perhaps the mutual support is permissible only if we wouldn't, from the nature of one of the hypotheses in question, expect some sort of independent observation, as we would here. Or perhaps there is *some* support in this case, just a trivial amount.

Still, the non-constructivist does owe us some additional answer here. Lakatos details a list of efforts to confirm independently that the exculpatory hypotheses are true, and we would certainly make these efforts. It is difficult to see, on the face of it, why the non-constructivist doesn't make fault tracing entirely too easy, and why that account of confirmation doesn't make the wrong prediction about whether we ought to rest content with these mutual cases of confirmation.

All the same, constructivism has to admit that nobody actually *gave up* Newton's theory, and if there was a problem with the data, shouldn't somebody have done so? This orbit was a known inconsistency with Newton's theory for a very long time, until General Relativity accounted for it. Yet Newton reigned supreme nonetheless. Doesn't this show that we don't, in fact, feel any great need to confirm independently some suggested escape?

Even given the (somewhat suspect) claim that no one saw this as a threat to Newtonian mechanics, it would not follow that Newton's theory didn't forbid anything. Constructivism can say this: The hypotheses of Newton's theory were by far the best confirmed of their day. Scientists spent a lot of time and effort trying to find a correction, which they could confirm to be correct, which could cope with Mercury. Newton's theory, along with other independently confirmed auxiliaries, *did* forbid Mercury's orbit so far as they could tell, and towards the end it was getting very difficult to see how to cope with the anomaly.

But the scientists could not simply abandon Newton's theory despite this. Scientists need at least to accept theories in order to design experiments (van Fraassen 1980, 73). (Or at least, they need to accept *some hypotheses* of some theories to do so.) And besides the need to accept some hypotheses in order to do science, the evidence in favor of some hypotheses is simply overwhelming. Scientists are entitled to say that, in spite of the fact that a set of hypotheses prohibits something that we do actually observe, the positive support from the observations is sufficiently overwhelming to justify accepting those hypotheses.

So scientists were forced into a very uncomfortable position. They had to hope, despite good evidence, that a solution to the Mercury difficulty would eventually be found. But they could not give up, and did not propose to give up, the claim that Mercury's orbit was, so far as they knew, forbidden by Newton's theory.

I believe there is some evidence for this feeling of discomfort. Simon Newcomb, for example, was repeatedly exercised in trying to figure out where the difficulty lay (1882, 474–477; 1906). By exhaustion of alternatives, he was eventually forced to attack the $1/r^2$ proportionality for the force of gravity. But he didn't seem at all happy about it, saying that it required "other proofs" before modifying Newton's law in this way (1906, 17). This requirement of independent confirmation to restore consonance is the mainstay of fault tracing, and the fact that Newcomb demanded it even when nothing else seemed at all capable of saving Newton is limited evidence in favor of constructivism. (Corliss (1979) contains a useful collection of other sources addressing the history of the concern with the orbit of Mercury.)

5 Examples of Fault Tracing

Constructivism brings empirical justifications to a finite terminus. Because it does so, it can account for the way in which our scientific practice includes independent justifications for a single hypothesis, and for the auxiliaries we use within a justification. By making use of these independent justifications, we can locate which hypothesis is at fault when a set of hypotheses clashes with the evidence.

Is this, though, enough to match the reasoning we go through when we actually do locate a fault? Is this part of our ordinary reasoning in the natural sciences and in less formal empirical reasoning? For if it is not, then supporters of the Quine-Duhem hypothesis will argue that it is an irrelevant sideshow, and that we both can and do hold onto hypotheses come what may, and do not use constructivism. To show the relevance of fault tracing to real science, we need to show that constructivism can cope with examples that are plausible cases of fault-tracing in ordinary scientific reasoning. Not only is the constructivist view a theoretical possibility by which we may evade Quine-Duhem, it is the way we actually reason in natural science and ordinary cases of empirical reasoning.

If successful, examples like these constitute an indirect, and defeasible, argument that empirical justifications have to end in the difference between what we have actually observed and what we haven't, and nowhere else. For it appears that we use observation, and nothing else, to discriminate exactly where and how to alter our theories. We cite observations to make the discriminations that we do, and we find nothing else satisfactory for doing so, and, in some cases, the verdict is inescapable.

I hope the current chapter will also show how subtle the technique can be in discriminating between, not only which hypothesis is at fault, but also (intuitively speaking) exactly how it is at fault, and precisely in what way it ought to be modified, along with other hypotheses.

5.1 First example: Polarized sunglasses

This is a simple, and I think plausible, example of the way in which investigators might actually proceed to discover why they do not observe what they expect.

5.1.1 First stage: Observations are inconsistent with beliefs

Sunglasses are often made with lenses that transmit only vertically polarized light. (From now on I will use the word 'polarized' to mean vertically polarized.) The advantage of these over ordinary darkened sunglasses is that they preferentially filter out glare. Much of the glare we experience is reflected from horizontal surfaces such as the roofs of cars or the surface of water. This reflected light contains a lot of horizontally polarized light. Vertically polarized lenses, therefore, block a great deal of it, because:

> H1. Polarized lenses will block nearly all light at roughly 90° to their plane of polarization (little or none can be detected by eye when the angle is roughly correct as judged by eye).

On a sunny day, a customer discovers that some polarized sunglasses have been marked down. He puts a pair on, and looks through a second pair he holds up to the light. He observes that, as 1 states, he can rotate the second pair so that their plane of polarization is at right angles to those he is wearing, and the lenses of the second pair appear jet black.

He recalls that:

> H2. Polarized lenses at approximately 45° permit some light to pass (enough to be easily detected by eye).

So he takes a third pair of sunglasses in his other hand and interposes them between the first two pairs. He rotates the third pair while holding the first and third at right angles to each other, until the third pair of sunglasses makes an angle of 45° to each of them. As he expects, more light passes through these three pairs of sunglasses at 45° than passed through the original two pairs at right angles.

In a spontaneous and playful mood, he replaces the third pair with a fourth at the same orientation. Much to his surprise, though, almost no light is transmitted this time.

This result is inconsistent with his beliefs of course. 1 & 2 together with these claims entail that light should be transmitted:

> H3. Sunglasses one, two, three and four are all polarized.

> H4. Sunglasses one, four and two were at 45° to each other.

So clearly at least one of hypotheses H1–H4 must be false (Fig. 11).

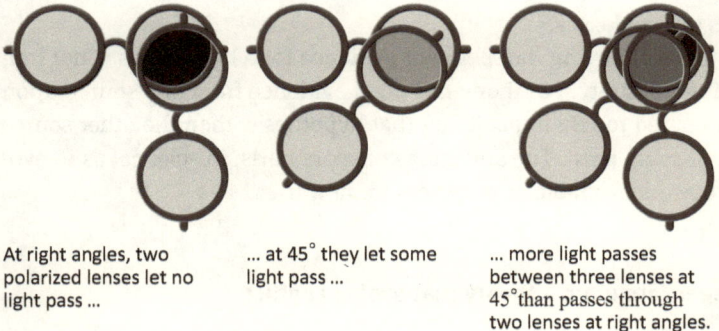

At right angles, two polarized lenses let no light pass ...

... at 45° they let some light pass ...

... more light passes between three lenses at 45° than passes through two lenses at right angles.

Figure 11: Properties of Polarized Sunglasses .

That is the first stage of the process of fault tracing. The customer knows of evidence that supports H1–H4, and in attempting to use them constructively to confirm 2, he finds a result he did not expect.

5.1.2 Second stage: Make an intelligent guess at where the fault lies

We need a hypothesis which will restore the consistency of the customer's beliefs with the evidence.

How does the customer know the lenses are polarized? He might have observed a sign on the box of sunglasses saying "Polarized sunglasses, on sale now!", or there might be a label on each of the lenses saying "polarized", or he might have been informed by the sales personnel, etc. Hypothesis 4 says that the sunglasses have been held at the correct angle. We can gauge this simply by observation.

Most people, I think, will trust physicists over the verisimilitude of salespeople. The customer is likely to have had experience with mis-labeled goods, either through incompetence or because some sort of profit motive is involved. So this is the most probable source of error. Do other things suggest to the customer that 1 and 2 are true?

The evidence for 1 (the properties of polarized lenses) comes both from his observation of the first two pairs of sunglasses, and from experiments done in physics class, and from statements of others who have proved trustworthy. The first two pairs of sunglasses blocked all light at right angles, so if one wants to retain the claim that all the lenses are polarized, one will have an instance of 1 by direct observation. (Here I depend on the idea that we can simply observe by eye when the pairs of sunglasses are approximately at right angles.) As I noted, the customer has also observed an instance of sentence two when holding the first three pairs of sunglasses.

The importance of having independent evidence for a hypothesis is not limited to increasing its surety. For if one has good evidence from one source upon which one is prepared to rely in justifying that hypothesis, then the other source is, so to speak, 'a spare part'. The advantage of spare parts, in science, as in evolution, is that they are available as resources for new uses.

5.1.3 Third stage: Formulate a theory that avoids conflict

Hence, if one is prepared to rely upon experiments in physics class and the word of the textbook authors alone as evidence for H1, then one has the evidence of the first and second pair of sunglasses at right angles available to justify this variation on H3:

H3 a) The first and second pair of sunglasses both have polarized lenses

from the evidence of the experiment of holding them at right angles. But note that one cannot do this and retain the experiment of the sunglasses at right angles as evidence for H1. The theory of confirmation is not constructive if one both depends upon H3 a) to confirm H1 and depends upon H1 to confirm H3 a). But there are plenty of additional reasons to be confident that polarized lenses block all light at right angles, for there is the word of educators, and perhaps the experience of other experiments.

One is relying on the physicists for hypothesis H2 as well. Making use of that, and the experiment with the first, third and second pairs of sunglasses in sequence, will constructively confirm:

H3 b) The third pair of sunglasses is polarized.

Then finally use the darkness observed with the first, fourth and second sunglasses at 45 degrees to confirm constructively:

H3 c) The fourth pair of sunglasses is not polarized.

Now we have a set of hypotheses, expressed by sentences {1,2,3 a-c, 4}, that is consistent with the data and which speculates that 3 is false. This process bears similarities to solving a murder mystery, where the writer has included an interlocking set of clues that point to a specific suspect.

5.1.4 Fourth stage: Confirming the diagnosis

To do this, we need additional evidence that constructively confirms that our suspect is actually responsible for the original problem. Hypothesis H1 has been strongly justified by physics experiments. So it can be used to constructively confirm H3 c) and thus restore agreement between the data and the hypotheses. One can, therefore, place the fourth pair of sunglasses at right angles to either of the first pairs and observe that some light is transmitted. One can also bolster this experiment if one uses other pieces of independently confirmed background knowledge. For example, one could rotate the fourth pair of sunglasses around and see if there is any change in the transmission of light reflected off the roof of a car. If H3c) is correct, then there should not be. One will also want to confirm H3b) by the same experiment.

Again, if one possesses other pieces of background knowledge, one will want to compare the relative merits of the various sunglasses at blocking glare, perhaps at different orientations. And one may want to bolster the evidence for having the sunglasses at the correct orientation by repeating the experiments with two and three sunglasses in sequence with various combinations of the sunglasses at various orientations to each other and to one's eye. One doesn't want to spend much time doing all this though, because by now the store will have telephoned the police.

5.2 Second example: Compton's description of fault tracing in measuring X-ray wavelengths

5.2.1 Background

What follows is a striking instance of fault-tracing uncovered in the process of measuring X-ray wavelengths, as detailed by Compton and Allison (1935).

In 1912, Max von Laue conceived of a way to measure X-ray wavelengths. We measure the wavelengths of visible light by using a diffraction grating. These can be either mirrors or transparent surfaces with very fine lines ruled on them. When coherent light is reflected or transmitted from these, the wavefronts form reinforcing maxima at an angle α to the incident waves according to the formula:

$$n\lambda = d \sin \alpha,$$

where λ is the wavelength of the light, d is the distance between the ruled lines, and n is a positive integer called the order of the maximum. This had all been very well established and thoroughly understood for the case of visible light in the nineteenth century.

The difficulty is that this formula will only work if d and λ are roughly as large as each other. For X-rays, λ is tiny, roughly the size of atoms, and as we rule finer and finer diffraction gratings, the limitations imposed by the size of atoms eventually set in. Laue's ingenious solution was to use the atoms themselves like the lines in a diffraction grating. If we have regularly spaced rows of them, the X-rays should be diffracted in a regular pattern. The solution is to use a crystal, and depend upon its regular organization of ions.

And one enormous advantage of crystals, which will figure prominently in the discussion that follows, is that we know how far apart the ions are in them. For we can weigh them, and use the molecular weight of the substance and Avogadro's number to get the number of ions, and the volume and shape of the crystal to infer their distance from each other.

Laue, therefore, suggested that we establish that X-rays had a wavelength by looking at whether there were maxima of intensity of X-rays scattered when a coherent beam passed through a crystal. The results confirmed exactly that.

5.2.2 Bragg diffraction

There is, though, a major difficulty. Ions in crystals are not parallel lines in two dimensions, but regularly arrayed points in three dimensions. While transmitting X-rays through a crystal produced maxima and minima, they were extremely complex and, at the time, unintelligible as a method for measuring wavelength.

William L. Bragg came up with another ingenious solution. Suppose we looked at X-rays "reflected" from a crystal, rather than transmitted through it, and measure the diffraction at some point where the angle of incidence of the X-rays (θ) exactly matches the angle of "reflection". (This isn't genuinely reflection, despite its analogy; the X-rays are being diffracted from the ions after they penetrate the crystal. At some angles, the waves reinforce to produce a maximum. At others they cancel to produce a minimum.) At these planes Bragg showed that the diffraction pattern should be particularly simple. The diffraction from the "horizontal" ions in the crystal exactly cancel each other, and the pattern is due only to the "vertical" planes of ions (Fig. 12).

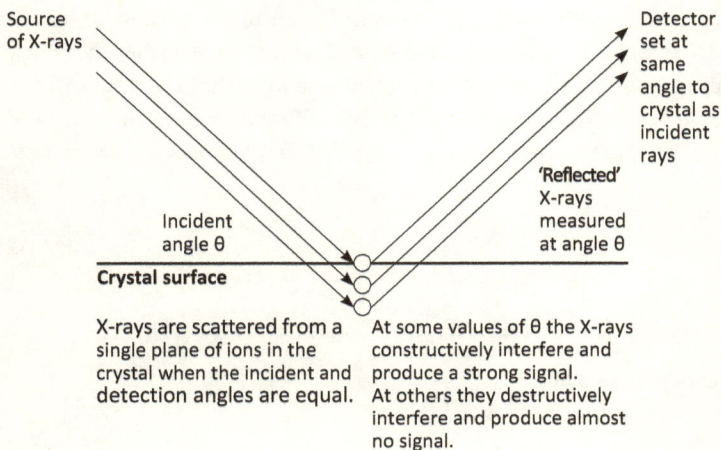

Figure 12: Bragg Diffraction.

Here the incident and "reflected" angles are both θ, where different values of θ produce maxima according to the Bragg equation (Compton & Allison 1935, 345):

$$N\lambda = 2\,d\sin\theta.$$

Here λ is the wavelength, N is a positive integer, the N-th maximum, θ the angle of incidence, and d is the spacing between the ions in the crystal. We can calculate d from our knowledge of the crystal structure (see below) and N and θ are directly observed. Now we can use a beam of coherent X-rays (generated by accelerating electrons into heavy metals) to estimate the wavelengths of X-rays.

We used an identical method to measure the wavelength of coherent visible light from maxima and minima generated by reflecting it from the regularly spaced lines ruled on a mirror diffraction grating. Compton checked the calculated wavelengths against different maxima and minima, and against crystals of different known interatomic distances. Different physical setups for the apparatus were also used, and different sources of coherent X-rays. By 1931 this method had generated values for the wavelengths of X-rays with errors of less than 0.1% (Compton & Allison 1935, 695).

5.2.3 The problem, and initial attempts at a solution

By yet another ingenious method, Compton managed to measure wavelengths from a ruled diffraction grating. The method depended upon the fact that the index of refraction for X-rays entering a crystal is slightly less than unity, so that at very small angles a beam is reflected from it as if it were a smooth object. By carefully ruling lines and then tilting the surface at a shallow angle to the incident beam, he was able to get a measurable diffraction pattern using the directly measured distances ruled in the grating. Figure 13 shows the setup.

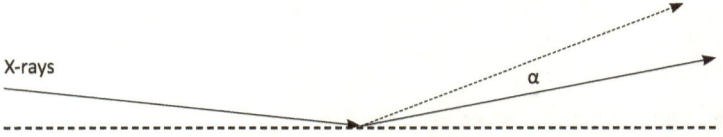

Figure 13: Interference at angle alpha.

Where a is the distance between the ruled lines, the angle α of the N-th maximum gives an estimate for the wavelength λ at angle of incidence θ of:

$a(\cos\theta - \cos(\theta + \alpha)) = N\lambda$ (Compton and Allison (1935), 691).

There are thus two methods for measuring the wavelengths of X-rays, which used different objects as a diffraction grating. One used the spacing of atoms in a crystal, and the other used lines ruled mechanically on a tilted surface.

Startlingly, these new values consistently differed from the evidence from the crystals by at least 0.15%. The difference reappeared with different sources of coherent X-rays and different experimenters. Although every value in the original calculation had been thought to have been known to a high precision, either one of these values, or the new method, must be in error.

5.2.4 Trying to discover the source of the error

One advantage of equations is that one can search through the parameters for the location of the error in the measurement. It is not at all easy to figure out what might have gone wrong. Compton went through an extended series of experiments, all of which failed to find the error.

First consider the errors that could arise from the ruled grating. The analysis of diffraction maxima and minima was well understood from other kinds of light, so that the difference could not lie there. X-rays could be detected by a variety of methods so that the angle of the incident beam could be very carefully evaluated. Similarly, the effect of its divergence could be carefully evaluated and shown to be inadequate (Compton notes a maximum value for this effect of a few thousandths of one per cent (Compton & Allison 1935, 697)).

Compton next checked the ruled grating and the engine used to make it very carefully. As a result of suggestions by others, he repeated the experiment with X-rays that covered a larger number of ruled lines, so that periodic errors in the line spacing could be averaged out. He also double checked for this with a variety of gratings under a variety of different circumstances. Systematically increasing irregularities in the distances between the lines will not account for the discrepancy either (Compton & Allison 1935, 693). Penetration of the material by the refracted X-rays, shadows cast by one ruling on its neighbors, and multiple scattering were also considered and rejected as potential causes of the difficulty. In each case, the explanation was inconsistent with results from other experiments, or was experimentally established to be in the wrong magnitude or direction. As a result of this exhaustive testing, the ruled grating was secure as a source of accurate measurements of the wavelengths of X-rays.

(By 1935 the experiment had been repeated by many people, with the same persistent discrepancy in the results. The ruled grating was always a slender reed on which to hang the results. The gratings were generated by a mechanical process, using devices that were extraordinarily thoroughly investigated and widely used. The number of lines per millimeter could be counted with a microscope (Compton & Allison 1935, 694) and the interference effect was well substantiated in the range of visible light for ruled gratings.)

The upshot of all this was that the ruled gratings were safe. Where, then, could the discrepancy be located in the experiments using the crystals? The equation from which the wavelengths were calculated depended upon three kinds of parameters; the angle of incidence of the beam (θ), the analysis of diffraction, and the spacing between the ions in the crystal (d). The first and second of these were again shared between the two analyses, and could again be double-checked by the same kind of experiments mentioned above. The interatomic spacing between the ions in the crystal lattice was therefore the likely source of the difficulty.

This spacing was calculated from the gross density of the crystal, and the number of atoms its mass contains. Let m be the mass of the crystal, M its molecular weight, V its volume, and A Avogadro's number:

d = number of atoms / V
Number of atoms = mass of all (m) / mass of one atom
Mass of one atom = M/A
So:
d = m A/M V

The puzzle is that all these seem to be too well established to account for the discrepancy too.

Perhaps crystals are more, or less, dense at their surfaces, in a phenomenon similar to surface tension in a liquid. Since the layers of ions near the surface of the crystal dominated the measured first order maximum, this might account for the result. Experiments at the time on higher order maxima, which must involve ions much deeper in the crystal, only served to confirm the observed discrepancy, however (Compton & Allison 1935, 698). Perhaps there were irregularities in the crystal structure, so that the gross density was in error. If there were any extra crevices or spaces in the crystal, though, this would give a discrepancy in the opposite direction to that observed.

Maybe the ions in the crystal overlap a bit, enough to account for the discrepancy. The trouble with that suggestion is that the spacing of the crystal could be measured by the data of X-ray diffraction itself, using only the theory of interference and refraction shared by the two methods, as well as known values of Avogadro's number, gross density, and the molecular weight of calcite. These measures substantiated the value for interatomic spacing that had previously been calculated, making it difficult to maintain that the discrepancy could be accounted for by overlaps in the crystal lattice. This test was repeated, with the same result, for a wide variety of crystals. The calculated value of the crystal spacing was the same, whether the supporting observations derived from:

1. The gross density, weighing the sample, Avogadro's Number and the molecular weight of the chemical, or instead from:
2. Angles of refraction, density, interference theory, Avogadro's number and the molecular weight of the chemical making up the crystal.

It is important that interatomic spacing was confirmed for crystals of varying chemical compositions. If there were an error in the measurements of gross density, then one would expect the different crystals to give varying values for the wavelengths of the X-rays, but this was not the case. Molecular weights could be eliminated from the list for a variety of reasons. Even wet bench chemistry had determined them to significant enough figures to rule them out as anything but a trivial contributor to the problem, and by 1930 the mass spectrometer gave far more precision.

The variety of crystals meant that many chemicals were involved for the same magnitude of discrepancy. Their molecular weights couldn't all be in error by the same amount in the same direction, for the obvious reason that molecular weights are relative measures. To say that the molecular weight of P is r is just to say that r grams of P combine with s grams of Q, or t grams of R, where s and t are the molecular weights of Q and R respectively. The different measurements of X-ray wavelengths for different crystals and the different calculations of the interatomic spacing show that the error must lie in the constant used to calculate them all, but not used in the ruled grating experiments: Avogadro's Number.

5.2.5 Restoring agreement with the data

Let me review. There were two experiments, using the same physical theory (diffraction gratings), giving different values for the wavelengths of X-rays generated in identical ways. The major difference between the two was the way in which the distance between the sources of diffracted waves was measured. In the ruled grating experiment, it was measured by using microscopes, and by well understood and independently confirmed theories of machines used to rule the grating. In the case of crystal lattices, it was calculated using different experiments, which nonetheless share values for some variables and constants. These were the macroscopic density of the crystals used, their molecular weight, and Avogadro's number.

All of these experiments had been revisited. The ruled diffraction grating seemed unimpeachable partly because of its long use in measuring the wavelength of visible light, which could be independently checked (for example, by using Young's slits). Since the calculations of the interatomic spacing in crystals were internally coherent, and could be checked by different kinds of experiments, it looked very much as though the error must lie in the physical constants and variables shared by these.

Avogadro's number had been calculated by dividing the charge needed to deposit one mole of a monovalent element by the charge on an electron. Charge could be precisely measured because current and time could be. The value of the charge on an electron was established by Milikan in his famous oil-drop experiment.

The best determinations of Avogadro's number came from electrolysis. We can measure precisely the current and time necessary to deposit a measured mass, and then use Milikan's value for the charge on an electron:

$$\text{Number of atoms} = \frac{\text{current time to deposit one mole}}{\text{charge on the electron}}$$

Again, current, time and mass could all be determined with greater accuracy than would account for the anomaly, so the likely candidate was the charge on the electron.

In Milikan's famous experiment, oil drops were observed to reach a terminal velocity v_g when falling under a gravitational field. An electric field of voltage V caused them to rise at terminal velocity v_e. The charge on the oil drop, q, was then:

$$q = K/V(v_g + v_e)$$

The constant K was:

$$K = 18\pi \, (\eta^3 \, v_g/2g \, (\rho_{oil} - \rho_{air}))^{1/2}$$

where η is the viscosity of air, g the acceleration due to gravity, and ρ the relevant density of the fluid.

In 1932, Kamekichi Shiba showed that Millican's value for the viscosity of air was based on a method that gave values that were consistently too low. When values that had been subsequently accepted as more accurate (entirely independently of the X-ray experiments) were substituted, the discrepancy between the two experiments to measure X-ray wavelengths disappeared (Compton & Allison 1932, 704; Shiba 1932).

For those familiar with them, the policy of locating which auxiliary to test by comparing different experiments that use overlapping sets of auxiliaries is very reminiscent of Mill's methods. It is some advantage to constructive confirmation, then, that it is broad enough to encompass these methods, since we obviously use them sometimes.

5.3 Third example: Tracing errors to the experimenter

5.3.1 Purposes of this section

Constructive confirmation does not deny that interests of the experimenter play a large role in what scientists do. It states only that this role is not decisive, and that observation can and does provide a way to avoid it. This section aims to illustrate this by showing how observations can sometimes locate the fault in an experiment by citing human error in the experimenter.

5.3.2 An interesting and simple example

This episode happened at my college, Mount Holyoke, which admits only women as students.[11] The (accidental) "experiment" was an attempt to teach the chi-squared

---

11 Mount Holyoke now also admits transgender students.

test to non-mathematics majors by integrating it with a historical episode, and consisted of a Monte Carlo experiment performed by the students. (I should acknowledge Robert Schwartz and the mathematics department at Mount Holyoke who developed the course materials and taught the course.)

In the Salem Witchcraft trials of 1692, 89 people were formally accused of witchcraft. Only some of the accused were executed in the ensuing trials, as shown by Table 2.

Table 2: Were female witches more likely to be executed in Salem?

Executed?	Female accused	Male accused
No	54 (74%)	19 (90%)
Yes	19 (26%)	2 (10%)

Does this result indicate that an accused was more likely to be executed if she was female?

On the face of it, the case looks quite convincing. Twenty six percent of the women were executed and only ten percent of the men, so it looks as if women were twice as likely to be executed. (Some accused were children, but I will say "men" and "women".) And the sample size of 89, while hardly large, wasn't small either. The students, in company with most people, I think, tended to believe that there was a good case for saying that accused women were more likely to be executed than accused men.

The chi-squared test answers this question: If accused women were no more nor less likely to be executed than accused men, how frequently would we observe a result which is at least as extreme as that which we actually observed? That is, assuming the 89 accused had the same probability of execution whether they were male or female, and that 68 women and 21 men are accused, and that 16 of the accused were executed, how probable is it that 14 or more of those executed would be women? It is a quantitative way to address the question that was addressed informally in the last paragraph.

A chi-Squared test doesn't by itself say anything about the probability of the hypothesis. Rather, it gives one a comparison of the probabilities of observing data of different sorts depending upon whether the hypothesis is true or false. The standard convention is to reject a hypothesis when and only when the data we have actually observed will occur no more frequently that in one in every twenty trials, if the hypothesis, and the background assumptions, are true. Deborah Mayo emphasizes the importance of this and other tests of the same nature (1996).

If accused women are just as likely to be executed as accused men, results at least this extreme would occur about one time in three. Most people, I think, find this a surprisingly high frequency, and are willing to say that the example provides much less evidence of bias than they had thought initially.

Rather than showing the (rather involved) mathematics to the students directly, they first did a Monte-Carlo simulation of the situation. The students were divided into small teams and put 89 beads in a bag, 68 round and 21 polyhedral to simulate the 68 women and 21 men accused. Then they randomly drew out (without replacement) 16 beads to simulate those executed, and recorded how many of each sort of bead had been withdrawn.

Pedagogically, the results were a disaster. The intention of the simulation was to teach the students that their informal judgments about the likelihood of bias were unreliable, that it wasn't all that unusual to see a distribution like this. When the results of the many teams were tallied, however, they departed significantly from the theoretical predictions. Specifically, there was a surfeit of results in which more men than expected were executed. The surfeit was in fact very marked, in which results such as those in the table hardly ever occurred.

5.3.3 Tracing the fault

What had gone wrong with the experiment the students performed? One might worry that the beads had been miscounted, and challenge the observations. Alternatively, we could challenge the auxiliary, and argue that the beads hadn't been randomly selected. We are backtracking through the constructive tree to get these hypotheses about what went wrong. Support for the idea that the selection of beads was not random came from observing the students themselves performing the experiment.

The students were aware of what they were simulating, and of course they were all women. They were selecting beads from a bag, and the "male" and "female" beads differed in their shape. When they extracted the beads, they were not neutral between whether the bead symbolized the execution of a man or a woman. The students would say things like "Yeah! Another man!" when they pulled out a polyhedral bead. Teams were disappointed or cheered depending upon the ratio of polyhedral to round beads that they "randomly" extracted.

The most promising line, then, was the suggestion that the selection wasn't random. The students need not have been deliberately attempting to get men executed (had it been a conscious effort they could simply have cheated, resulting in a far more biased sample). But we are all aware of ways in which desires have far more subtle effects upon behavior. The beads were different shapes, so that a student could have detected by touch whether a "man" or "woman" would be pulled out.

The background knowledge that experimenters are often influenced to get the result they want, coupled with what the students were observed to say, supported the hypothesis that they were not randomly selecting the beads, but that a desire for a certain kind of result had influenced their behavior. This was the intelligent guess at what had gone wrong that the instructors made.

My colleagues in mathematics confirmed that this was what had happened by making the beads different colors, but the same shape, so that unconscious bias could not affect the selection. After that, the results coincided with the frequencies one would expect for the chi-squared test.

5.4 Fault-tracing in natural science

5.4.1 Synopsis of fault-tracing

In the first stage of the process, we get a conflict between a set of hypotheses, $\{H0\}$ and a set of observations $\{E0\}$. There are constructive trees refuting $\{H0\}$, perhaps in conflict with each other about some element of that set.

To make an intelligent guess about where the fault lies, one uses the information one possesses. One could, first, depend upon the fact that $\{E0\}$ contains multiple independent justifications of only some hypotheses in $\{H0\}$, and guess those are not at fault. Or one might introduce more information about what outcomes of observation have occurred or are likely given other information one possesses, $\{E1\}$. Or one might combine the approaches. The objective is to identify hypotheses $h \in \{H0\}$ that are better justified, and hypotheses $h^* \in \{H0\}$ that are worse justified.

Then attempt to formulate a set of hypotheses, $\{H1\}$, which are all constructively confirmed by $\{E0\}$ and by any additional outcomes one has used to discriminate the justified and unjustified hypotheses among $\{H0\}$ (that is, the elements of $\{E1\}$). This is the third stage of fault-tracing.

The above processes depend upon a number of constructive trees. $\{H1\}$ in effect states that a subset of $\{H0\}$, call it $\{h\}$, are acceptable, and a disjoint subset, $\{h^*\}$, are to be rejected. Seek to confirm that each h and $\neg h^*$ really are correctly accepted and rejected by seeking constructive justifications of them that are independent of the trees one has used to make the diagnosis.

That is the fourth stage: restoring consonance with the data. The problem arises because we have a set of trees, confirming not all elements of {*H0*} and refuting at least some of them. To restore consonance, we are seeking new trees confirming {*H1*}. We'd like these to be evidence independent of the elements of {*E0*} and hypothesis independent of the elements of {*H0*}. That would reassure us that we weren't depending upon something that gave rise to the original problem, and was therefore suspect. So one is seeking new evidence, and new ways to confirm that the correction is the right one. This is at least suggestive of the things philosophers have written about the importance of novel tests in natural science.

5.4.2 Closing words: The reply to reductionism and verificationism

Fault-tracing is a normal and widespread part of science. A suspect hypothesis may be found innocent, as well as guilty, when we test whether it is at fault. The test it faces will be independent of at least some other hypotheses which formerly made us confident or doubtful of it. Fault-tracing can trace guilt, or innocence, to individual hypotheses, quite independently of our present evidence for or against them, and sometimes independently of the company they keep in gaining that present justification. When we know that a hypothesis is a candidate for future fault-tracing, we know that that hypothesis can run a risk, and gain a victory, independently of its fellows. We know this even when we know that it can never be tested alone.

The purely relative view of confirmation has a much more difficult time in accounting for these features of fault-tracing. Purely relative confirmation holds that the observations never bear upon a single hypothesis, but only collections of them. It would seem that making more observations never can show that a hypothesis is more likely to be the innocent than the guilty one when the observations are not as we expected. For, to arrive at that conclusion, we would have to know how probable it is that the subsequent observations were due to the innocence of the hypothesis we are trying to test, or instead the guilt of some other hypotheses that we depended upon in this subsequent experiment. But to get this piece of knowledge is exactly the problem that the subsequent experiment is supposed to be solving – we are trying to find out which of two possible flaws is the real one.

I cannot see a way to solve this problem if confirmation is purely relative. But I observe that we do solve this problem, using the outcomes of observation, all the time in our scientific practice. I conclude that our practice cannot only be based on relative confirmation, and that it must be constructive, at least in those cases where we are able to tell what's gone wrong. Those cases are very far from being unusual.

In "Two Dogmas" Quine faulted his opponents for embracing what he termed 'The Verification Theory and Reductionism' (1953 [1980], 37–42). I will close by briefly pointing out explicitly something that must have been obvious for some time, at least to those familiar with epistemology. Fault-tracing, while it is inconsistent with the Quine-Duhem hypothesis, is just as inconsistent with verification theory and reductionism.

As Quine intended the terms, verificationism was the view that ". . . the meaning of a statement is the method of empirically confirming or infirming it" (1953 [1980], 37). Reductionism was the same position with an explicit reference to experience (38–39). Both views hold that what we believe, when we believe some hypothesis, is that the evidence will come out in a way that confirms it and not in a way that refutes it. This is the difference that adopting the belief will make as compared to not doing so; people who adopt will expect the evidence (whether understood subjectively or not) to support the belief.

It's essential that some form of *translation* should be the aim. That is, the statement of the hypothesis must be capable of being reformulated, at least in principle, as a statement about expectations concerning things we can observe. The simplest case is one Quine called 'radical reductionism'. "Every meaningful statement is held to be translatable into a statement (true or false) about immediate experience" (38). Other versions of reductionism or verificationism might alter the idea that immediate experience was the way to capture observation, but could not deviate from the aim of translation. The intention was to avoid reference to entities viewed as suspect, and to keep to something observable.

This is wholly incompatible with any scientists realizing that hypotheses are subject to fault-tracing. For if fault-tracing is a part of science, then *we cannot at any time know* what observations will eventually turn out to be the ones that we will take to confirm any particular belief. It depends upon how the processes of fault-tracing will go in the future. The whole point of that process is that *we do not know what will eventually be confirmed or refuted*. We have to *actually go through* the process of doing new experiments and seeing what happens to see what turns out to confirm or refute any hypothesis which we now believe. Depending upon what happens, we will try new experiments and see what happens there.

Think of Compton's experiments. Suppose Compton adopted the belief that the ruled grating was going to give the true value of X-ray wavelengths at the beginning of his experiments, before the conflict with the crystals had even been discovered. He could not have anticipated *how* that belief would eventually be vindicated. It depends upon the course of future observation, and our ingenuity in reasoning from a sequence of new observations, and the new experiments they suggested. There is no way to list the range of possibilities by which Compton's belief would turn out to be justified. We have no idea of that range, and cannot have. All we can say is "I expect that observations will turn out in a way that confirms my belief". We cannot so much as conceive of the limits of this range at the time we adopt the belief, let alone translate the statement of the belief into them.

Constructivism denies the Quine-Duhem hypothesis. But it does not, and cannot, do so by adopting reductionism.

6 Chang's Paradox

Hasok Chang (2004) gave a detailed argument that certain identified hypotheses cannot be constructively confirmed. The argument is striking for its historical fidelity and detail. Chang argues that hypotheses he displays cannot be justified by the outcomes that we have observed at all, never mind constructively. In this chapter, we will look at one of these examples, and how it could, indeed, be constructively confirmed.

The example is instructive for another reason too. One need not expect a theoretical proposal of the way in which hypotheses are justified in natural science to follow history in every detail. One need only look at the history of calculus to see an example of a practice that was adopted and its claims believed before its justification was presented in a defensible way. But if a proposed model of justification need not follow history slavishly, it could still be refuted if it deviates too much. Historical scientists cannot regard hypotheses as firmly and unproblematically justified when the model says the justification is unavailable to them.

So we ought to be able to see how the justifications that constructivism proposes are available to natural scientists at the times that they regard hypotheses as settled. They need not have thought of matters in exactly the way constructivism says, but a reconstruction of the justification after the fact ought to look roughly right, and to qualify reasonably as the sort of justification that could have been offered and ought to have looked convincing.

6.1 Chang's paradox and its solution

6.1.1 The paradox

Hasok Chang writes:

1. We want to measure quantity X.
2. Quantity X is not directly observable, so we infer it from another quantity Y, which is directly observable. [...]
3. For this inference we need a law that expresses X as a function of Y, as follows: $X = f(Y)$.
4. The form of this function f cannot be discovered or tested empirically, because that would involve knowing the values of both Y and X, and X is the unknown variable that we are trying to measure. (2001, 251; 2004, 59)

I've called this *Chang's paradox*, although Chang himself calls it "the problem of nomic measurement". Chang argues, in effect, that we cannot constructively confirm that the reading on an instrument measures an unobservable quantity, because one has to introduce the way the observable relates to the unobservable by fiat, as an assumption that cannot have been supported by the outcomes of observation.

Look at step 4 – the step I will challenge. If we had some other means to measure X, there would be no problem. The issue arises because it appears impossible to acquire initially some means of measurement because to do so requires some antecedent method. So this is another original acquisition problem: it claims that the evidence cannot give you the function f in the first instance. Chang's example concerns the original design of thermometers using changes of volume as a way to measure temperature more accurately. The observations cannot possess the resources to associate volume (observable) to temperature (unobservable) by a linear relation. If we associate them that way, it cannot be the observations that favor it over, for example, a logarithmic relationship.

A closely related paradox is called the *experimenter's regress* by Harry Collins (1985, 2, 84; 2004, 128–130). Here again, the puzzle is how we could build a detector for something we think might be there, but haven't yet detected (gravity waves in Collins' extended example). One detector detects the entity, and another doesn't. How do we know which detector is better? Well, the better detector is the one that gets the correct result. Which is the correct result, detection or absence? Obviously, we should believe what the better detector says about whether the entity is there. But we can decide which detector is better only if we know whether the gravity waves are really there. And we can know whether the gravity waves are there only if we can decide which is the better detector.

Again we have an unobservable: gravity waves. The problem concerns the first success at detecting these. In order to know that we possess some means to detect them, we require some other method of detecting them, to establish that the detector really works.

The comparison of different measurement procedures for the same unobservable is really a variant on Chang's paradox. Suppose we are given two measurement procedures A and B that are supposed to measure a single quantity X. Call the observable outcomes of the two procedures Y_A and Y_B. We think that B and A both measure quantity X, that is, that some function of the observables, $f_B(Y_B)$ and $f_A(Y_A)$, both give the value of X at a time and place. Now how could we possibly establish that A and B measure the same quantity independently of establishing that f_A and f_B are the correct functions in each case? Only if they are the correct functions will the agreement in outcomes about the value of X provide some evidence that they're measuring the same quantity. But the converse holds too; only if we assume that they're measuring the same quantity will we be able to use one of them to establish that the other function is correct.

The constructive answer to cases like these is that we do not need some independent access to the unobservable X to show that $X = f(Y)$. Rather, we observe some features to support first some very weak hypotheses about X and Y, and then by building up more and more observations of the features of the directly observable world, we use the weaker hypotheses to eventually justify the $f(Y)$ function.

6.1.2 Initial steps in measuring temperature

Chang's paradox is, in effect, a challenge to confirm independently two hypotheses:

A: Fluids (for example, mercury) expand linearly with temperature.

B: We possess devices (for example, mercury thermometers) that measure temperature.

Neither A nor B can be confirmed without using the other. So it is impossible to confirm constructively that we have measured a temperature.

But we can constructively confirm B (that mercury thermometers measure temperature) independently of A (that mercury expands linearly with temperature) by drawing parallels between measuring the lengths of objects, and measuring the temperatures of samples of different ratios of volumes of boiling and iced water (if we have length, clearly, we can get volume).

First, we need observable things to try to measure, analogous to the birthday-cards in the example of length. So produce samples of boiling to iced water with different ratios of volumes; 10:0, 9:1, 8:2, etc. We do not need to assume at the outset that all 9:1 samples (for example) are the same temperature. We must begin with some justification for believing that we can observe that something is a 9:1 sample, but that isn't difficult if we have volume.

As Chang so ably relates, boiling and freezing are not simple matters (2004, 8–39). They are affected by the purity of the water, the atmospheric pressure, and the cleanliness of the vessel and the presence of dust. They are processes that develop over time, not instantaneous and repeatable like length. So it appears that, once again, we cannot even begin to justify hypotheses about temperature without assuming a great deal about temperature, and we have cycles of confirmation, or empirical assumptions that cannot be empirically justified.

As Chang also relates, though, it is pretty easy to bring about circumstances in which water and these primitive thermometers repeatedly bring about consistent results so far as gross observation is concerned (Chang 2004, 11–39, 48–56). There was a long struggle to identify features that affect boiling (and freezing). The purity of the water matters, as does the pressure of the air, the presence of irregularities in the vessel, and dust in the air. Chang points out that these features were often identified only after thermometers were available (2004, 11–39). Is our constructive justification of *B*, then, doomed to circularity?

In the case of temperature, unlike length, we have to do a great deal of fault-tracing right from the start. Chang relates that doubts about variation in boiling and freezing points were evident right from the beginning of historical investigations (2004, 11–17). Many of these difficulties were observable with the naked eye. Food takes noticeably longer to cook at altitude, and salt melts ice. Bear in mind that we are dealing with an original acquisition problem. We will begin with *some* systematic observable set of circumstances in which we can justify that boiling and freezing are occurring at roughly the same temperature. At least with the naked eye, these circumstances produce boiling and freezing in predictable ways. We can leave the rest to mopping-up operations later on.

Unlike Chang, I do not hold that the human body's sense of temperature bears some special relationship to the justification of thermometers (Chang 2004, 43). Suppose you had no sense of heat. You could still experiment on samples of various ratios of boiling and freezing water and observe the behavior of mercury thermometers. Even if the whole human race had no unaided sense of temperature, we could still build thermometers and justify using them.

6.1.3 Establish that mercury thermometers measure temperatures

The next steps are closely analogous to the process by which we established that rulers measure length. I will only repeat the outline briefly.

We have two mercury thermometers, $T1$ and $T2$, and a variety of samples of different ratios of boiling and ice-water. Just as we used the two rulers on disjoint sets of cards, use each thermometer on disjoint sets of ratios. (That is, use $T1$ on ratios 10:0, 8:2, 6:4, etc. and $T2$ on 9:1, 7:3, 5:5, etc.) Observe that on each sample of a particular ratio, $T1$ comes to the same (marked) point. Similarly for $T2$. Use these instances to confirm, of each thermometer, that:

> If Tn is immersed in two samples of water of ratio $o:p$, (Tn will reach point x both times if and only if each sample is at temperature y).

And now cross-check the results between the two thermometers in the same way you did for the two rulers (that is, use *T1* on the samples you formerly used for *T2*). Establish that:

The temperature of sample 1, ratio 9:1 = the temperature of sample 2, ratio 9:1.

Repeat to show that every sample of water of ratio 9:1 has the same temperature, and similarly for ratios 8:2, etc.

It is wise to extend the justification to new cases. For example, one can take a sample of alcohol and show that one thermometer always reaches the same mark whenever it boils. Infer that the other will repeatedly reach an identical mark made on it too. Test this. Build a new mercury thermometer and predict that it will give consistent results like the first two. Try boiling oils or freezing salt-water.

What you're doing is showing that, under certain independently identifiable circumstances, observable events like freezing, boiling, or the ratio of composition, are at constant temperature. Then you use predictions based upon generalizations from these events to predict that all mercury thermometers will approximate the same mark at the same temperatures. Any of these experiments could fail.

And indeed, as Chang points out, they do fail to agree if you're not very careful indeed. Herman Boerhaave asked Daniel Fahrenheit to make him two identical thermometers. But he found that the thermometers did not quite agree with each other. They were made at the same time and place, by the same (skilled) artisan, of the same design, and calibrated by the same techniques using the same fixed points.

Fahrenheit was at a loss to explain the matter, but noted that the thermometers had been constructed of different kinds of glass. Perhaps different kinds of glass expand at different rates. Boerhaave accepted this as probably being the fault. But consonance wasn't restored until much later. The episode occurred in 1732 (Chang 2004, 57–8). It wasn't until 1847 that Henri Regnault examined the matter carefully and confirmed that different samples of glass expanded at different rates for the same change in temperature. Not only the type of glass, but even the treatments it had undergone affected its rate of expansion (Chang 2004, 79).

It might seem inadequate to begin to establish measuring instruments by making copies of something and comparing the results under various observable conditions, as I have suggested that we do for the ruler and the thermometer. Chang gives a detailed example, though, of scientists attempting to establish the reliability of various designs by doing the same thing. Regnault's experiments compared many designs of thermometer against each other and selected as the best those that could agree with each other and which gave consistent measurements under the same circumstances (Chang 2004, 74–84).

(I hope it is obvious that I'm not proposing that the *meaning* of the measured quantity is somehow exhausted by these agreements in operations. It is an essential part of constructivism that consonance can only be restored by independent tests, and that measurement of some quantity can be improved. The best of these will be unforeseen ways of "getting at" some quantity, or testing some hypothesis. So the significance of assertions of hypotheses concerning some quantity cannot be exhausted by operations of measurement, because that would prevent future, novel, methods from measuring a physical quantity that we know about now.)

Now remember Chang's original challenge. We need to confirm B independently of A:

A: Fluids (for example, mercury) expand linearly with temperature.

B: We possess devices (for example, mercury thermometers) that measure temperature.

I think we have done this. For we have confirmed B without supposing any hypothesis whatsoever about how the volumes of mercury will be related to one another. The experiments I've described will work if mercury expands as the log of temperature for example. They will work if mercury shrinks with increasing temperature. All we know is that mercury reaches the same mark at the same temperature, and some other mark at some other temperature, to within known approximations. This is all we need for B.

Once we have B, A is less easy than one might imagine. The volume of mercury at the temperatures of the ratios immediately suggests a linear scale. The distance between the marks on the thermometers for the 0:10 ratio and the 1:9 ratio is the same as the distance between the 9:1 and 10:0 ratios, so the mercury must have expanded by as much. It doesn't force this scale upon us, though. Perhaps the *difference* in temperatures between 0:10 and 1:9 ratios is different from the temperature difference between 9:1 and 10:0. I will return to this in a moment.

I would make the same kind of reply to Collins' experimenter's regress as I have made to Chang's paradox. There are methods for establishing that a detector is reliable and unreliable that are independent of whether what it is supposed to be detecting is really there. Indeed, this is exactly what happened in the subject Collins addresses, Joseph Weber's gravity-wave experiments (Weber, e.g., 1960, 1969, 1973, 1974, 1975).[12] Garwin and Levine pointed out, with considerable

---

[12] Weber tried to detect energy deposited from gravity waves in large bars of metal. The work is now widely regarded as poor. Collins 2004, especially chapter 9, describes it.

asperity, that Weber's device had failed in both directions (1973). It had failed to detect things that certainly were there, because they'd been put there by independently established methods, and it had detected things that couldn't possibly have been there. (Weber "detected" a signal moving at the speed of light between two points on earth by noticing a spike in the signals separated as one would expect if the signal moved between them at the speed of light. He failed to notice that the data he used were from two different time-zones.).

6.2 Ordinal and linear scales

I said I would return to the issue of whether there was as much difference in temperature between 0 and 10°C and 90 and 100°C. Bradford Skow argued that there is no evidence for this prior to the invention of, at least, statistical mechanics (Scow 2011, 472). Hasok Chang is more cautious, and suggests that heat capacity and energy might provide such a justification (2004, 41, 65). I side with Chang.

The point of this section is to show that there is a justification, long predating statistical mechanics, for thinking that the Centigrade scale gets the temperature differences correct. It is therefore quite reasonable for early scientists to adopt a Centigrade scale, or others that are linear transformations of it. The point of showing *that*, in turn, is to emphasize what I hope is already apparent. We can constructively confirm much more than might at first appear. It isn't unreasonable to suspect that whatever we can really confirm, we can confirm constructively. (This justification for the centigrade scale is only approximate. Chang relates a historical episode that was a failed attempt to fault-trace its imprecision (2004, 170).)

Skow gives an account, common in many physics books, of one way to establish a scale of temperature (Scow 2011, 477). We take two fixed points, which we get from standard procedures like ice-water and boiling water, and then we divide the difference between them into a conventionally chosen number of units, ("We *decree* that the number measuring the system's temperature on this scale is equal to 100 d1/d2, where d1 is the distance between the 0 mark and p, and d2 is the distance between the 0 mark and the 100 mark" (emphasis added, Skow 2011, 477).) A skeptical student could well demand: "Yes, but why divide it up into *equal* parts? Why say that these numbers *register the value* of the temperature? Why not say, instead, that it just tells whether things are hotter or colder than one another?"

S. Stevens first raised this distinction (Stevens 1946, 678–679). Some scales simply rank substances in terms of whether they have more or less of some quantity. We can rank the hardness of minerals by seeing which ones scratch which other ones, and assigning increasing numbers. This is an *ordinal* scale, and it only indicates, by the number it assigns, whether a mineral is more or less hard than another, not how much harder. What we want is a *linear* or *metric* scale, in which, as Scow puts it, the temperature difference between the sun and his study is 34 times the difference between his tea and his study (473).

In the case of length, there is an intuitively appealing experiment one can do to compare lengths at different magnitudes, namely, move rulers around and lay them end to end. We could define rulers on a strict analogy with the way Scow suggests we define a temperature scale. We could take an object, or observable class of objects – say, the class of things that can be brought into coincidence with the foot of some king – decree that it is to be one foot, and divide it into twelve equal units. But nobody believes that length is an ordinal scale only. Why not? Because the obvious methods of comparison have observable results that support a metric scale instead. What we need is a similar, constructively confirmed method of comparison for temperature.

6.2.1 The course of experiments

What we want to do is establish that temperature (unobservable, at least at this point), and volume of thermometric fluid (observable, given that length is so thoroughly confirmed) vary as a linear function of each other, not as some other function. To do that, we need some other means of characterizing temperature changes that we can confirm without supposing that any particular function relates volume and temperature.

To do this, I've chosen energy-transfer as the independent method of characterizing temperature. I'll confirm constructively that the same changes in temperature are accompanied by the same transfers of energy between observable objects. Then I can use the quantity of energy to induce a change in temperature, without supposing any particular relationship between volume of a thermometric fluid and temperature change. We will observe, when we do so at different temperatures, a constant change in the volume of thermometric fluid, that is, a fixed change in the length of the column of mercury in the mercury thermometer. So volume and temperature are linearly related.

So far, we have certain resources. In particular, we know that mercury thermometers at least measure an *ordinal* scale. So a thermometer reaches the same point on its scale whenever it is immersed in fluids at the same temperature, and rises and falls as temperature goes up or down, although we do not know by how much. We can, with a little work, get fluids to boil and freeze at constant temperatures too (or at least, they appear constant using naked-eye tests).

We know how to measure weight. Take an object made of copper, weigh it, and let it sit in boiling water for a while. Now move it to 1 liter of liquid water at the same temperature as ice. Observe the change to a thermometer at equilibrium. Repeat. The temperature change is always the same with identical weights of copper. Then repeat with other bodies of copper of different weights to show that weight really makes a difference. Repeat with iron.

Hypothesize that something is being conserved, energy. Guess that this thing seems to be present in a certain amount for each kilogram[13] of metal immersed in boiling water. Energy gets transferred to the ice-water from the metal, a fixed quantity per kilogram, which varies with which kind of metal we use. Confirm all this by calculating what weights of which metals are needed to raise the water temperature to the *same* temperature, as measured by a mercury thermometer. (An ordinal scale will give us a way to establish when temperatures are the same, of course.)

We know that 1Kg of copper from boiling water always imparts the same temperature change to 1 liter of water at the temperature of ice. We can repeat the experiment by immersing the copper into boiling oil too. We find that that, too, results in a constant rise in the temperature of 1 liter of water at the temperature of ice. So far, though, we cannot compare the magnitude of these two temperature changes.

If our theory about the transfer of energy is correct, though, we can calculate what weight of iron at the boiling point of oil will produce the same change in temperature to 1 liter of ice-water as did 1Kg of copper at the boiling point of oil. This prediction could fail. The amounts of energy absorbed by 1Kg of a metal might vary at different changes of temperatures, so it is perfectly possible that they would vary at considerably different rates for different metals. In fact, at these temperatures, they vary only very slightly, too small to detect with crude instruments, and the prediction succeeds (at least roughly). So now we have some evidence that the same transfer of energy results in the same change in temperature.

Plainly, we are going through the same rigmarole as we did for the balance. Changes in energy and temperature are hypothesized to be related by a single equation:

$$\Delta E = cm\Delta T$$

where c is the specific heat of the substance to which energy is transferred, m its mass, and ΔT the resulting change in temperature. In the case of the balance, we could establish when two points were at the same distance from the armature

---

[13] I know. A kilogram is a unit of mass, and we can only measure weight. Until we get some dynamics, we will have to settle for forces. We can still get started though.

without knowing the form of the equation that relates distance and weight. Then we could use that to confirm the equation which our theory proposed. In this case, we can establish when we have the same change in temperature and compare different observable solids and fluids to establish that c is a constant.

There are obvious ways to get additional support. Observe that different volumes of ice water attain different temperatures with the same pieces of metal from the boiling water. Hypothesize that the water absorbs a fixed quantity of energy per liter from them, and test this with different metals. Do the same experiments by immersing the metals in alcohol at the freezing point of water. Use the hot, or cold, lumps of metal of known weights to "carry" energy around.

Now calculate what weight of copper at ice-water temperature will absorb a known quantity of energy from water at boiling point. Put that weight of copper in freezing water, and transfer it to that volume of water at boiling point, and observe the temperature drop. That weight of copper will (we have already confirmed) transfer the same energy, and yield the same temperature change, to the same volume of water at freezing point. Compare the changes in length of the column of mercury in both cases. They are identical (to within the approximations the instruments allow, which are admittedly quite large). You've now "compared" the interval between 0° and warmer water, resulting from a donation of energy, with the interval between 100° and cooled water, resulting from the absorbtion of energy. The same change in temperature, as measured by energy transfer, yields the same change in volume of the thermometric fluid. The function relating the two must be linear.

6.2.2 What must be taken for granted?

Chang argues that calibrating a thermometer required two hypotheses. *A*, above, is that a mercury thermometer's length indicates temperature, and *B* is that mercury expanded linearly with temperature. *A* can only be confirmed using *B* as an assumption, and *B* can only be confirmed using *A* as an assumption. Assume something different concerning *A*, that it's the log of length that indicates temperature, and we measure a different result about how mercury expands. Assume that mercury expands according to some different relation, and we get a different result about how to calibrate thermometers. So mutual justification is inevitable.

Not so, I argued. We can use one set of observations to show that the mercury always reaches the same point at the same temperature, and justify only an ordinal scale for the thermometer. Then we can use that hypothesis and experiments concerning heat capacity to compare differences in temperatures.

Chang goes on to use the example to argue that scientists depend upon what he calls 'ontological principles' (2004, 91). These are assumptions, essential to the justifications that natural science presents, that are neither justified by logic nor by the outcomes of observation (2004, 91; 2001, 12). They are regarded as "essential features of reality in the relevant epistemic community" (2001, 11). They are very reminiscent of Quine's cultural posits in "Two Dogmas" (1953 [1980], 44).

An example of one of these is the *principle of single value* (Chang 2004, 90; 2001, 11). Each real physical quantity has just one exact value. Chang cites the fact that Regnault compared a great many thermometers, of various designs, to establish which gave the most accurate readings. Those thermometers that agree as closely as possible with themselves, upon repetition, and with others of the same design under identical circumstances, were identified as the best (2004, 60–89, 90). Regnault's tests didn't depend upon other hypotheses concerning temperature, such as the theory of mixtures and specific heats, but only compared the readings of thermometers at different times, and with other designs (2004, 89).

If temperature is a vague quantity, or mercury thermometers are measuring a different kind of temperature from, for example, gas thermometers, then Regnault's experiments will not yield the result that was so universally accepted. Yet this assumption – single valuedness – was needed to proceed with the empirical justifications, not justified by them (2004, 91). The outcomes of observation cannot govern which hypotheses we ought and ought not to believe by themselves. Science history would make no sense under such a monarchy, and it would be impossible to practice science.

Against Chang, I argue that the physicists of the time had observed outcomes that did indeed empirically justify the principle of single value. As a result, they had a good reason to take the most self-consistent designs to thermometers to be the best. Chang himself describes these experiments and their outcomes. The argument is fallible – empirical justifications always are. The observations made, though, had the outcomes that supported single-valuedness. Other outcomes, ones we did not observe, support a vague quantity, or more than one quantity. Because past scientists did not observe these other outcomes, they had reason to believe in the principle of single value.

There is, most prominently, the fact that we can get better and better agreement between different occasions on which we measure the temperature of boiling water (Chang 2004, 11–39). We can exclude more and more reasons for variation between the samples and establish that these are the reasons for variation. Chang very ably describes, for example, the way in which scientists discovered that irregularities in the vessel promoted much more consistent results concerning a fixed boiling point (2004, 17–28). If it is possible to discover observable features which, when varied, promote more consistent measurement outcomes, then that suggests a single value. Even the ability to form a rough-and-ready kind of scale is very limited evidence in favor of a single value.

Compare the concept of a species to that of temperature. This is a genuinely vague scientific concept "measured" by observed phenotype. There are certainly variations of phenotype within a single species across space – two species of gulls in northern Europe turn out to be overlapping "ends" of a single interbreeding continuum that circumscribes the North Pole. American bird-watchers will be familiar with other examples of regional variants (for example, in orioles and flickers). But controlling for space has no effect. When one moves different sub-species to new locations, they do not change phenotype to that which is present at the new location. So the different locations are not sources of error in measurement for which we can control to reveal a single, non-vague, underlying species-concept. The point a mercury thermometer registers as the boiling point varies with air pressure too. But when we make repeated measurements at the same pressure, the observed variation shrinks markedly. We can get more and more consistency in what we observe about temperature as we fault-trace. What counts as one species resists sharpening in the way that we can sharpen the ways we detect temperature.

I would argue that other outcomes of observation might have bifurcated temperature too. Alcohol and mercury thermometers, even when calibrated at fixed points, vary over the interval between them (Chang 2004, 58). If, for example, one of these had correlated with some phenomena in a linear way (for example, speeds of chemical reactions), and the other correlated with a disjoint class of phenomena in a linear way (for example, mixing of fluids of different temperatures), we might have justified some theory under which there were two different forms of temperature. (Recall that we once failed to distinguish weight and mass, for example.)

Once again, it's helpful to compare the case to a different example. Mass features in two distinguishable ways in Newtonian mechanics; in the phenomena of inertia, including the second law, and in the law of gravitation. These two epistemically possible kinds of mass must be proportional if Newton's theory was correct. This was recognized immediately, and a series of experiments established that, if they were not proportional, the variation must be extremely small (the most famous is by Eötvös in 1885). There would be no point in performing these experiments if they could not have had outcomes that were at odds with Newton's view that we are dealing with one physical quantity here. Natural scientists are aware of the need to establish that physical quantities are unitary, and do experiments to establish this.

Thus I would argue that Chang overstates his case when he says that "[a]ny reports of observations that violate the principle of single value will be rejected as unintelligible [. . .] this principle is utterly untestable by observation" (2004, 91). It is true that in a normal context, the demand for this kind of justification is bizarre.

Perhaps it is also true that there is a bias, in natural science, against raising such a challenge. But it doesn't follow, nor is it true, that we cannot respond to such a challenge if it gets raised. We could display experiments, with reasoning, to give at least some justification here.

Even if successful, the argument above doesn't show that there are no hypotheses that we are required to believe without argument if we hold that outcomes of observation deliver knowledge. Perhaps induction, or the hypothesis that I am not a brain in a vat, qualify. Still, it's important that the outcomes of observation give us a wider range of justifications than at first appears, for that suggests that we have much less ability to hold on to any hypothesis no matter what outcomes we observe.

7 Escaping Cycles of Confirmation

Constructivism gives identity conditions for a single justification – a constructive tree. Different examples confirming a single hypothesis can provide independent justification for it, so that we have more than one justification for a single hypothesis. So we can use the evidence itself to show that that target is likely true even under the hypothesis that one of the auxiliaries in one of the trees is false. The outcomes of observation need not, then, automatically allow us to avoid the target hypothesis if we reject a particular auxiliary.

To be constructive, though, each such tree must avoid cycles of justification. H1 cannot be justified from the evidence using H2 if the only way to justify H2 is to depend upon the truth (or probability, or other justifying virtue) of H1. Such a situation may occur, and be significant, but it is at best a provisional justification, requiring supplementation. It must be a legitimate part of ordinary scientific practice to reject the proffered justification.

Under the model of justification given by Quine-Duhem, though, cycles of justification are perfectly legitimate. We ought to observe convincing examples of them all over scientific reasoning. If we can find even one such example, where the constructive analysis is plausibly unavailable, and skepticism is unreasonable, then that presents an enormous difficulty for constructivism.

I have only been able to locate a single example where an author has argued that a cycle constitutes real confirmation just by itself. This is, once again, an example from Clark Glymour (1980, 178–203). Instead of having a cycle of two hypotheses, as in the last paragraph, Glymour presented a historical example of a cycle with three. He points out three hypotheses from Copernican astronomy, each of which was confirmed from the same data-set if the other two were taken for granted. Following Glymour, I present this data-set as a table at the beginning of the first main section of this chapter.

I show in this chapter that we can break out of the cycle Glymour presents, and could do so as soon as Copernicus presented it. Not only did each of the three hypotheses have evidence that favored it, but Copernicus actually gave this evidence in the same place where he describes the data that Glymour uses to give his example. So not only is this cycle one we can break out of, it is one that Copernicus successfully broke out of as soon as he presented it, just as constructivism suggests he should.

Copernicus' theory, in fact, contains a great deal to illustrate the kinds of justification that constructivism recommends. Ptolemy, by contrast, does contain at least one inescapable cycle, one that did indeed provoke skepticism. I argue at the end of this chapter that this might have been one of the things that recommended Copernicus over Ptolemy, even before the telescope was invented. The harmoniousness that Copernicus thought recommended his theory over its (then) empirically equivalent rival turns out under constructive analysis to connect genuinely to truth. Far from being a counterexample to constructivism, the attractiveness of the Copernican over the Ptolemaic systems, prior the invention of the telescope, illustrates the fruitfulness of constructivism as compared to the Quine-Duhem account.

7.1 A cycle in Copernicanism

Glymour presented a table of results (1980, 185), originally from Ptolemy (Table 3).

Table 3: Observed relationships between superior planets.

	Cycles of anomaly	Revolutions of longitude	Solar years
Mars	37	42	79
Jupiter	65	6	71
Saturn	57	2	59

For each planet, the sum of the first two columns is equal to the entry in the third column. The column titles are names of astronomical phenomena given by the Ptolemaic system. They are observations, made from the viewpoint of the earth, originally on the assumption that it is a stationary body.

A *solar year* is the most familiar. It is the time between two successive cycles in the pattern of paths of the sun through the sky. Say, the time between noon on two successive midsummer days, when the sun attains its highest point above the horizon.

Revolutions of longitude and *cycles of anomaly* require a longer explanation. Every body in the sky appears to execute a circle around the Pole Star, Polaris, once per day. Most of the stars visible in the night sky – the fixed stars – retain their positions with respect to each other, and appeared in fixed groups, or constellations. In addition to moving across the sky every night, different constellations are visible at different times of year. We appear to be at the center of a vast sphere, which rotates once a day from east to west, and once a year from west to east, so that different parts of it are visible at nighttime in different months.

By mapping with respect to the fixed stars, we can observe that, over the course of a year, the sun moves against the background they provide, through a circle called the *ecliptic*. The circle the ecliptic makes through the stars is not quite where the "equator" would be if we take Polaris as the north pole, but is inclined at about 23 ½ degrees to it.

Not all stars retain their positions with respect to each other. The inferior planets, Mercury and Venus, are never very far from the sun, and travel with it around the ecliptic every year. The superior planets, Mars, Jupiter and Saturn, also stay fairly near the ecliptic, and over many solar years, make a complete, somewhat erratic, circuit around it. A *revolution of longitude* is the time each superior planet takes to travel this circuit. Saturn takes longest, Jupiter next, and Mars takes the shortest amount of time.

I said that the motions of the superior planets were erratic. Their average motion is from west to east. This average motion is interrupted by periods when they move from east to west – retrograde motion. These are observed when the sun is directly opposite the planet when we locate it in the sphere of fixed stars. The period between two retrograde motions is a *cycle of anomaly* (Fig. 14).

So the table means that Mars, for example, moves westward 37 times in 79 solar years. During that time, it undergoes 42 revolutions of longitude – that is, it makes 42 orbits against the fixed stars. Obviously 37 + 42 = 79, and the same sum works for each of the superior planets.

Under the Ptolemaic system, this result is simply a remarkably striking coincidence. Revolutions of longitude are explained one way, and cycles of anomaly by another mechanism that need have no connection to that explanation. But Copernicus makes identifications that make these results inevitable.

Under Copernicus, of course, the sun is at the center of the solar system,[14] and we get the following identifications:
1. The solar year is the period of the earth's orbit around the sun.
2. The revolution of longitude is the period of the planet's orbit around the sun.
3. The period of a cycle of anomaly of a superior planet is the interval between two successive collineations of the sun, the earth, and the planet. (That is, the earth "overtakes" the planet in its orbit – see Fig 14.)

Glymour points out that, when any two of these hypotheses are taken as auxiliaries, the data confirm the third.

---

14 Here, as elsewhere, I'm speaking loosely to avoid unilluminating complications. Copernicus was aware that the sun was not at the center of a circle around which the earth orbited. All the same, their relative motion is very nearly circular.

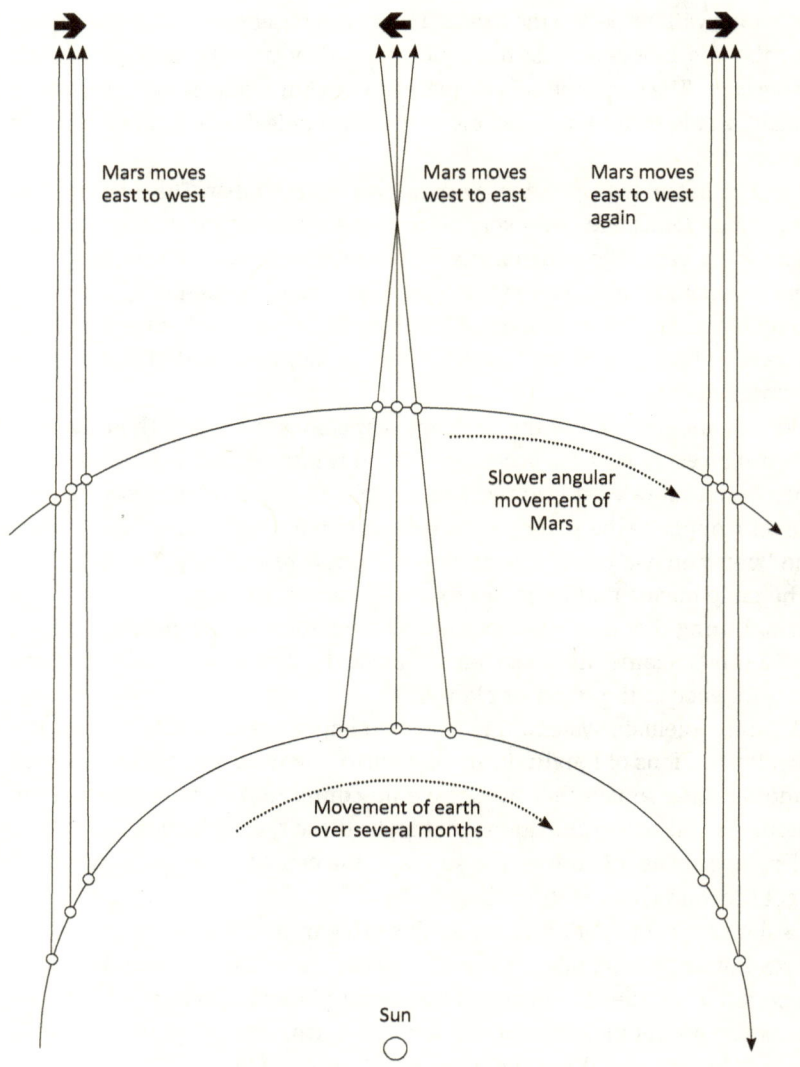

Figure 14: Retrograde motion under Copernicanism.

Suppose, for example, that we consider the orbit of Mars, and assume hypotheses 1 and 2. In 79 orbits around the sun, the earth is in line between Mars and the sun 37 times. Given this, the idea that we observe one revolution of longitude every time Mars orbits the sun predicts 42 revolutions of longitude, which is what we observe. So hypothesis 3 is confirmed. (If, in 79 laps around the track, a fast car overtakes a slow one 37 times, then the slower car must have made 42 laps of the track in that time.) Exactly analogous reasoning means we confirm 2 for the data if we assume 1 and 3, and 1 from the data if we assume 2 and 3.

All this is apparently genuine empirical justification. Intuitively, we *ought* to be convinced of all three of these hypotheses given this reasoning. Yet this is a cycle of justification. None of the three hypotheses has been traced back through a constructive tree. So it would appear that constructivism cannot be a necessary condition for empirical justification, since this is an example of non-Constructive, but intuitively genuine, justification from the evidence.

Constructivism does say that, just by itself, this cycle of justifications gives us no reason to believe any of the three hypotheses. But this cycle doesn't exist by itself. There are reasons for believing hypotheses 1, 2, and 3 that are constructive. So we can break out of the cycle, and we could do so in 1543.

7.1.1 Copernicus had independent constructive evidence for each of the three hypotheses

The fact that Copernicus himself proceeds in a constructive manner strongly suggests that the intuitive desire to produce constructive justifications informed the kinds of arguments Copernicus presented, even if it is a kind of inarticulate motivation of which he was not fully aware. He presents the resources for constructive reasoning, indeed, in the same context as this example (at 9^b of *De Revolutionibus* (henceforth, DR), Copernicus 2000 [1543], 25).

Hypothesis 1: A solar year is the period of the earth's orbit around the sun
As one would expect for this most important of the innovations of Copernicus, he puts the case for it at the beginning of Book one of DR ($3^a - 4^a$, 2000, 12–13). The argument consists only in showing that the question is open. Once we have noticed that day and night can be referred to the movement of the earth, rather than the whole universe around the earth ". . . it would not be surprising if someone attributed some other movement to the earth in addition to the daily revolution" (DR 3b; 2000, 13). The daily revolution from East to West of the entire Universe could be a daily rotation of the earth on its axis from West to East. And the annual revolution of the sun through the fixed stars from West to East might equally well be a movement of the earth around the sun in a clockwise direction as viewed from the North Star's perspective. We can observe that the sun moves in a close approximation to a circle with respect to us, but that

observation alone doesn't show which of these two bodies is stationary (with respect to the fixed stars, as we would now say).

The obvious reaction to this is that it is a very poor justification. Copernicus has not shown that there is a reason to prefer to attribute motion to the earth rather than the sun; he has only shown that this epistemic possibility is not excluded by the observations. But I would argue that that it is not such an insignificant step to raise the probability of the earth's movement to one half. Nobody else of that time put the subjective probability anywhere near as high (to speak anachronistically). Copernicus pointed out the relative motions of these bodies. These motions are at least *compatible* with heliocentrism, which made that hypothesis more probable than formerly for most astronomers of the day. I will argue below that this initial, impoverished, justification will rise when it can be independently justified.

Copernicus does use some background knowledge. The evidence is the apparently circular annual motion of the sun, and the constant size of its disc (thus, the orbit is of a fixed radius). These correlate exactly with the solar year, as determined by either the change in the seasons, or as measured by day lengths, or the height of the sun at noon. At this date, astronomers must have used the fixed stars to determine the circular annual, as opposed to diurnal, motion of the sun. (Other methods are possible, but observing constant motion around the ecliptic is the obvious solution.) If one includes this, it's also a premise that the fixed stars do not rotate around the sun, which is confirmed by the identity of the solar and sidereal years. It is also a premise that the diurnal rotation of the earth does not precess, as indicated by the North Star's fixed direction. Hence the constructive tree (Fig. 15).

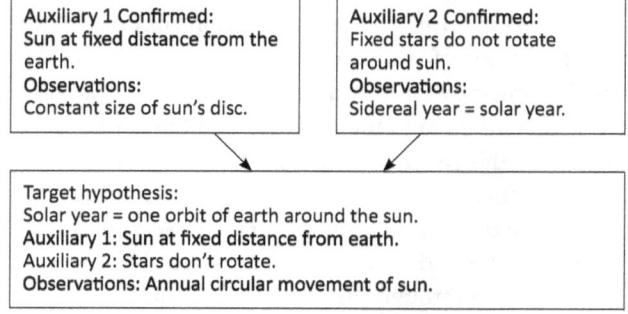

Figure 15: A Solar Year is one Earth Orbit.

Hypothesis 2: A Revolution of longitude is the orbit of a superior planet around the sun

Copernicus also points out that the planets approach and recede from the earth, as is evident from their relative brightness at different times (DR 3^b). He argued that the earth cannot be near the center of planetary motion because of this variation in brightness (DR 3^b).

Copernicus first gives arguments that Mercury and Venus in fact orbit the sun (DR 8^a-8^b). He then raises the epistemic possibility that the sun is also the center of the circles on which the superior planets cycle: ". . . if anyone should take this as an occasion to refer Saturn, Jupiter and Mars also to this same center, provided he understands the magnitude of those orbital circles to be such as to comprehend and encircle the earth remaining within them, he would not be in error" (DR 8^b).

He then gives the evidence favoring this possibility:

> For it is manifest that the planets are always nearer the Earth at the time of their evening rising, i.e. when they are opposite to the sun and the Earth is in the middle between them and the sun. But they are farthest away from the Earth at the time of their evening setting, i.e., when they are occulted in the neighborhood of the sun, namely, when we have the sun between them and the Earth. All that shows clearly enough that their center is more directly related to the sun." (DR 8^b)[15]

The superior planets are brightest when they become visible on the eastern horizon just as the sun sets, that is, when the Earth is in a direct line between them and the sun. They are faintest at those times of year when they become very briefly visible just prior to sunrise, that is, when the sun is almost between the Earth and them. So it is more probable that they orbit the sun.

Now given that they orbit the sun, their observed rotation once around the fixed stars must be identical to one circuit of the sun. For we know independently that the fixed stars cannot be rotating with respect to the sun. (Whether the earth goes around the sun annually, or the sun around the earth, rotation

---

[15] Note that the argument still works if the sun orbits the earth rather than conversely. Hypothesis 3 can be constructively confirmed independently of hypothesis 1.

of the fixed stars would be evident from the difference between solar and sidereal years.) We have the following constructive tree confirming hypothesis 2 (Fig. 16):

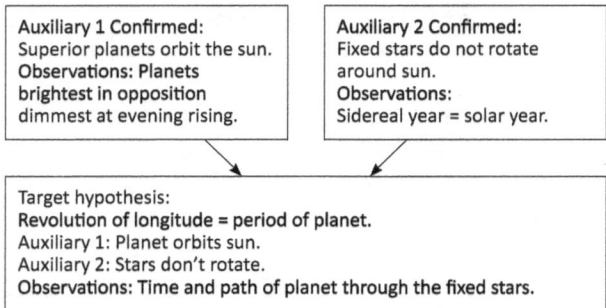

Figure 16: Revolutions of Longitude.

Hypothesis 3: A cycle of anomaly is the period between two successive collineations of the sun-earth-superior planet

While Copernicus doesn't directly state a justification for this, it has an obvious justification. That obvious justification is simply that we observe retrograde motion when and only when the earth is between the sun and the superior planet. Over a long period of time, equally obviously, the number of times when the sun-earth-superior planet are in line in that order is equal to the number of cycles of anomaly. Copernicus does make points about the observed magnitude of retrograde motion that are incompatible with his being ignorant of this justification (DR, 10ª). It could well be that Copernicus didn't state this justification because it was obvious to every astronomer of his day. Kuhn notes that knowledge of this relationship predates even Ptolemy by centuries (1957, 49–50).

We use the fixed stars, in this reasoning, as permanent locations on a sort of map. We see the planet move from east to west (retrograde motion) by observing its position over many weeks against this background. Without such a map, the motion would be much more difficult to spot because of the diurnal rotation of the whole system. It's technically possible to confirm hypothesis 2 without reference to the fixed stars, but I'll ignore that. We get a rather complicated single node justification (Fig. 17).

> Target hypothesis:
> Cycle of anomaly = period between colineations.
> Observations: Motions of the sun, planet and fixed stars.
> Retrograde motion just when sun in opposition to the planet.

Figure 17: Retrograde Motion.

Therefore, each of the three hypotheses can be confirmed constructively.

7.1.2 An independent justification for heliocentrism

We now have justifications for the three hypotheses that Glymour thought were confirmed in a cycle. Each of these were available to Copernicus at the time, so we are not forced to allow this as a convincing example of inescapable mutual confirmation.

When we looked at hypothesis 1 – that the earth's year was its orbit around the sun – we saw that Copernicus had very little evidence for it. The sun could just as easily orbit around the earth. We can now reinforce this rather meager justification. For we did not use the data of the table in any of the justifications for the hypotheses, and hypotheses 2 and 3 got confirmed without depending upon the truth of 1. We get a constructive tree with the following skeleton (Fig. 18):

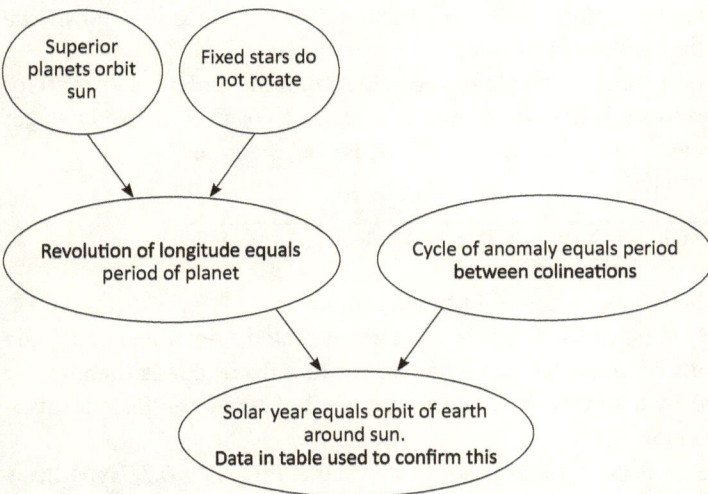

Figure 18: Constructive Confirmation of Copernicanism.

7.1.3 Amplifying justifications

We now have a constructive justification for both hypotheses 2 and 3. These two justifications are independent of each other, from separate pieces of evidence. The justification for 2 makes it very secure, for the sake of having a number, to assign a subjective probability of 0.95. The justification for 3 is more doubtful, because perhaps the variations in brightness are not due to distance. So give it a lower probability, say around 0.75.

I have already noted that the evidence for hypothesis 1 from the relative motions of sun and earth is very poor. Let's then set its probability at no more than 0.5. Nonetheless, because hypothesis 1 follows from the data in the table, plus the truth of 2 and 3, this poor showing receives a boost from its independent, and constructive, justification.

To be explicit, proceed in steps:

1. A Copernican contemplates the relative motions of the sun and the earth. As a result, he sets the probability of hypothesis 1, that the earth orbits the sun, at 0.5. This is one justification for hypothesis 1, and I now move to an independent one.
2. The Copernican contemplates the evidence for hypotheses 2 and 3. These are justified independently of each other, and independently of step 1. The motions of the planets with respect to the earth and sun could be as hypotheses 2 and 3 say, even if the earth obeyed very strange motions with respect to the sun. He sets *Pr(2)=0.95* and *Pr(3)=0.75*. Round down the probability that both are true to 0.7.
3. Now our Copernican contemplates the evidence in the table. Let T stand for this. What we want is the new value for *Pr(1)* given everything we have got so far. We need a value for *Pr(1/T^2^3)*. By Bayes' theorem:

$$\frac{\Pr(T|1 \wedge 2 \wedge 3) \times \Pr(1 \wedge 2 \wedge 3)}{\Pr(T|1 \wedge 2 \wedge 3) \times \Pr(1 \wedge 2 \wedge 3) + \Pr(T|\neg 1 \wedge 2 \wedge 3) \times \Pr(\neg 1 \wedge 2 \wedge 3)}.$$

4. I argue for the following values in this equation:
    a) *Pr(T/1^2^3)* = 1, or very close. As already noted, given geometry, the number of circuits executed by a fast car in a fixed time is the number executed by a slower car, plus the number of times the fast car overtakes the slower.
    b) *Pr(1^2^3)* = 0.35, or roughly so. At this stage, *Pr(2^3)* = 0.7. Hypothesis 1 is probabilistically independent of this, and I set its prior at 0.5. Multiplying, we get 0.35.
    c) On the face of it, *Pr(T/¬1^2^3)* is very low, close to zero. For if a solar year is not the earth's orbit around the sun, it's simply very difficult to imagine how the values in the table could come about. Even so, constructivism should be generous, as lack of imagination is not a strong argument. Take this probability to be 0.5. The values of the table are not *more* likely if the earth doesn't orbit the sun.
    d) *Pr(¬1^2^3)* = 0.35. The reasoning is the same as in case b).

5. In the light of this reasoning, the new probability that the earth orbits the sun is about 2/3, which is the approximate results when the values in step 4 are put in Bayes' theorem in step 3. If we are less generous about 4 c), the posterior probability is higher.
6. So, our Copernican has two justifications, the one he found in step 1, and the one developed in steps 2–4, above. These two justifications are independent. Steps 2–4 are consistent with just about any relative motions of the sun and earth, so long as the sun sometimes is between the two, and sometimes the earth between the planet and the sun. These observed relative motions of the sun and earth are what we used in step 1. Moreover, the observed relative motions of sun and earth are compatible with all kinds of observations and hypotheses about the planets and fixed stars.
7. From 6, then, we have two independent methods for confirming that a solar year is the orbit of the earth around the sun. Each raises the probability of hypothesis 1 whether or not the hypotheses used in the other justification are true. So the subjective probability that the solar year is one earth orbit should be higher with both than it is with each alone, since the prior probability that one of them fails to justify is independent of the posterior probability conferred by the other on hypothesis 1.

Independent justification, then, amplifies justification under constructivism.

7.2 Other ways that Copernicanism illustrates the superiority of constructivism

There are known cycles of "justification" in Ptolemy that were regarded with great suspicion and for which Copernicanism provided an escape to the observations. Consider a major puzzle for Ptolemaic astronomy. Mercury and Venus are never observed very far from the sun. They each orbit epicycles centered on a line between the earth and the sun.

Now is Mercury or Venus closer to earth (see schematics in Fig. 19 and Fig. 20.)? The center of the epicycles of both are tied to the sun; that is, the line from the earth to the sun passes through the center of both epicycles (the dash line in the diagrams). Is Mercury closer to the earth than Venus, as in the first diagram, or Venus closer, as in the second? Both present the same relative positions of Mercury, Venus and Sun as observed from the earth. Both capture exactly the same observations from the earth.

136 — 7 Escaping Cycles of Confirmation

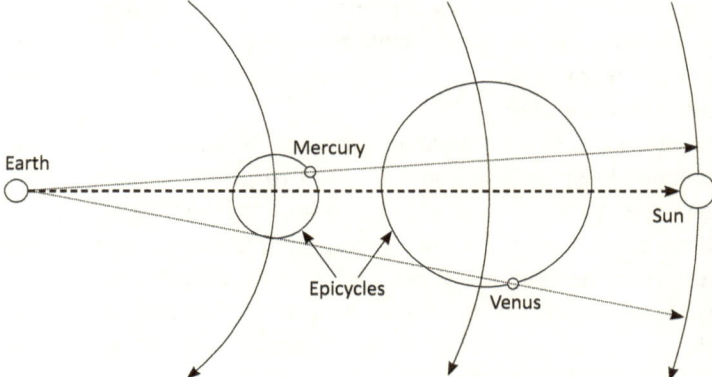

Mercury is closer to the earth than Venus,
note the angles between the dotted lines.

Figure 19: Ptolemy; Mercury closer to us than Venus.

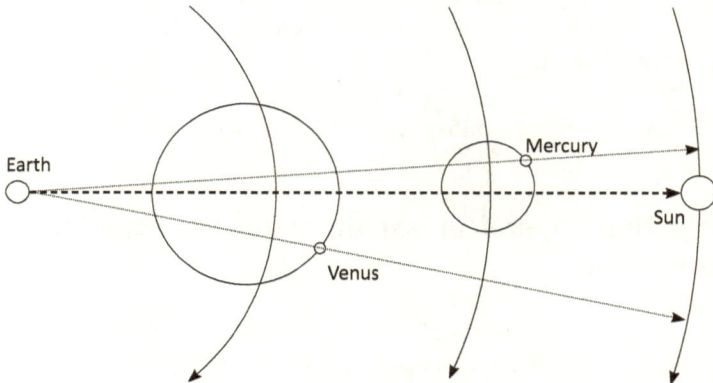

Venus is closer to the earth than Mercury,
note that the planets are observed at the same
angles to the sun when seen from the earth.

Figure 20: Ptolemy; Venus closer to us than Mercury.

Although the order of these inferior planets cannot be determined from the data, we can determine the ratio of their radii of the epicycles by using their maximum elongation from the sun. So consider the following two hypotheses:

(i) The farther an inferior planet is from the earth, the greater the radius of the epicycle.

(ii) Venus is farther from the earth than Mercury.

Using the greater maximum elongation of Venus from the sun, we can confirm (i) using (ii). Similarly, using the same data, we can confirm (ii) using (i). If Mercury had had the greater maximum elongation, we'd have a refutation in each case. Yet no-one thought that this settled the question of the order of the inferior planets, for the obvious reason that we could reverse the order just as well by reversing what (i) says (DR 7^b; Copernicus 2002 (1543), 20–21).

Copernicus mentions an attempt by Ptolemaic astronomers to break out of this cycle. Using observation, we can calculate the sizes of the epicycles of Mercury and Venus. If Mercury is nearer the earth, then these epicycles almost exactly 'fit' into the distance between the orbit of the moon and the sun. This is at least suggestive. If we suppose that all parts of the distance between the moon's orbit and the sun's must be included in the orbit of something, then there is evidence for Mercury orbiting nearer to the earth than Venus does. But we are again stuck in a cycle, as there's no independent justification for this supposition. There is no evidence, that is, that epicycles need to fill the space between the orbits of the moon and sun. Copernicus rejects the argument for this reason – ". . . we do not know that this great space contains anything except air or [. . .] the fiery element" (DR 8^a). It looks as though Copernicus, too, wanted to break out of cycles.

7.2.1 Other instances of constructive confirmation in Copernicus

The unobserved parallax of the fixed stars

Copernicus is aware that if the earth orbits the sun, then we would expect to observe parallax in the fixed stars. (That is: we would see them from different angles at different times of year, so their relative positions would not be constant.) Yet they clearly retain the same relative positions at all different times of year. This is some evidence that the earth is stationary, unless the fixed stars are so far away that parallax would be unobservable. So Copernicus presents an independent argument that they must be very distant from the earth, as compared to the planets. At so great a distance, we would be unable to detect any parallax.

Compare Cancer and Capricorn, which are at 180° to each other. When Cancer is first visible on the horizon in the evening (that is, rising), Capricorn is briefly visible in the morning (that is, setting). Six months later, we observe Capricorn to rise, and Cancer is setting at 180° to it, so the angle between the two remains the same. But, even if the earth is stationary, we should be observing these two collections of stars at quite different angles. For Cancer rises at a different point on the horizon from Capricorn, relative to an observer fixed at earth, and does not set at 180° to that point. Yet we observe no parallax. So, at least compared to the diameter of the earth, the fixed stars must be very distant indeed. Even if the earth is stationary, the opponents of heliocentrism have to concede that the fixed stars are much farther away than the planets, so that observable parallax isn't likely.

For my purposes the example is instructive. Copernicus is defending his case from an objection (DR 4^a-4^b). He's doing what we saw Darwin doing earlier, replying to an objection by showing that there is evidence throwing doubt on one of the auxiliaries that the objection appeals to.

The order of the superior planets

Copernicus endorses the view, ubiquitous throughout the prior history of astronomy, that we know the order of the superior planets (Mars, Jupiter and Saturn) by the apparent speed of their movements through the sky (DR 7^b, 9^a). Because Saturn changes position relative to the fixed stars more slowly than Jupiter, that is evidence that it is farther away. We would now regard this as doubtful, although it does offer some justification. The angular speed of moving objects nearer to an observer is usually faster, although this is not a sure guide.

Copernicus offers an additional reason for this order, based upon his account of retrograde motion (DR 10^a, 141^a). This account is defective by today's standards, but it follows the same qualitative structure as ours. Retrograde motion is due to the fact that as we approach the planet at a high angle to its motion around the sun, we observe more of its relative change in position with respect to the sun and fixed stars (see Fig. 14 again). As we "overtake" it our own greater angular speed around the sun dominates, and it "recedes behind" us, apparently traveling backwards. When we view its motion at almost a right angle again, after we have "overtaken" it, we again perceive its motion with respect to the sun and fixed stars.

If this is even qualitatively correct, we would expect to see the nearest planet, Mars, to move in retrograde motion more quickly than the farther planets. We ought also to notice that its progressive, west-to-east motion relative to the fixed stars is faster. That is what we do observe. Copernicus notes an additional point. We have observed Saturn moves more slowly relative to the fixed stars than Jupiter, and Jupiter more slowly than Mars. The various "overtakings" of the planets ought to be more frequent for Saturn than they are for Jupiter, and more frequent for Jupiter than Mars. So the cycles of anomaly ought to be more frequent too, and that is what we observe. There is no similar line of reasoning that links the progressive and retrograde motions of the superior planets in Ptolemy's system.

Copernicus is offering two pieces of reasoning, both of which confirm the order of the superior planets. These pieces of reasoning agree in the order they ascribe. The first says that objects nearer the sun move faster than those farther away. The second describes qualitatively the retrograde motion of nearer and farther planets. This second piece of reasoning shares with the first the auxiliary that farther objects apparently move more slowly, so the first is not independent of the second, although the second is independent of the first (recall that independence of justification is not symmetric under constructivism). There is no *a priori* reason why the data should justify the order attributed by the second of these methods, given only the outcomes on which the first depends. But they do agree. That is an independent justification for the order of the planets.

7.2.2 It is very difficult to square Copernicus with different data

Suppose a planet did not obey the relationship in the table. Suppose Mars (for example) had executed not 42 but 47 revolutions of longitude in 79 years, with the same 37 cycles of anomaly.

There is a standard view of what Copernicus would have done in that situation that stems from the Quine-Duhem view of justification. He would have used Duhem's dodge of altering the background knowledge. Copernicus could have saved his view by saying that the fixed stars rotate about the sun 5 times in 79 years, in the opposite sense to the rotation of the planets about the sun. Because a revolution of longitude is a circuit of the planet around the sphere of fixed stars, if that sphere rotates, then the planet will appear to have executed more rotations around the sun than it actually has. We will just modify hypothesis 3, and say that a mean planetary year is the difference between the rotation of the stars and the orbit of the planet.

Hence, the objection runs, the proclaimed instance of constructivism that Copernicanism allegedly provides is just a sham. Constructivism says that Copernicus tightly constrains the data in the table, and if they come out otherwise, then Copernicus is refuted. Nonsense. If the data had been refuting, Copernicanism would have just juggled the background assumptions in the way that Quine-Duhem says is always possible. So constructivism has no advantage over Quine-Duhem in this case.

But this analysis will not work. Suppose we try the Duhem dodge above with the planet Mars to evade the force of the counterexample. If Copernicanism is to be saved in the light of this observation by the rotation of the fixed stars, then the fixed stars must rotate (roughly) 1/16 of a circle every year.

But this would refute the data for Jupiter and Saturn. We count one revolution of longitude as the time it takes for a superior planet to reappear at the same latitude on the background of the fixed stars. Whence it follows that in the 71 years in which we observe Jupiter to execute 6 revolutions of longitude, it should in fact execute (roughly) 10 1/2 of them (because the background is rotating). But 10 1/2, plus 65 cycles of anomaly, no longer equals 71 solar years. Saturn, in its 59 years for 2 revolutions of longitude, ought to execute a little under 6 instead. In that time, it still makes 57 cycles of anomaly, since the cycles of anomaly are unaffected by the rotating background. But 57 plus 6 doesn't equal 59.

Well, maybe we can fudge that too, somehow. Perhaps Mars orbits the earth as a second moon. Nope. If Mars orbited the earth, then the idea that cycles of anomaly are "overtakings" has to go since Mars, unlike the moon, has cycles of anomaly.

But musings like this are really beside the point as far as trying to save the hypotheses in the light of the counterfactual evidence is concerned. For we have an *independent* method for checking whether the fixed stars rotate around the earth with a given angular velocity. We can check on the difference between a solar year and a sidereal year. Hence we are foiled at the first step; we cannot simply suppose that the zodiac rotates.

Well, a solar year is measured by the cycles of the sun's altitude in the sky. Perhaps the diurnal rotation of the earth precesses, so that we get more (or less) than one summer per annual circuit of the sun. But that won't work either, because then the angle to the North Star, and all other stars, would vary.

It is, in short, very difficult indeed to see how to save anything like Copernicus' view even given very small changes in a small subset of data. Because we can determine the values of different features of the whole system using different datasets, and because those values are so interdependent, different data than those we actually observe refutes the whole system. Conversely, if altered small subsets of data present very serious problems for a theory, then that theory must require that many different ways of determining its important hypotheses must agree. For if not it would be easy to alter a small part of the theory to cope with the altered data.

7.2.3 The advantage of the Copernican system

Copernicus wrote of his view that the earth and the other planets orbit the sun that:

> . . .not only do all their phenomena follow from that but also this correlation binds together so closely the order and magnitudes of all the planets and of their spheres or orbital circles and the heavens themselves that nothing can be shifted around in any part of them without disrupting the remaining parts and the universe as a whole. (2002, 5)

> ... in this ordering we find that the world has a wonderful commensurability and that there is a sure bond of harmony for the movement and magnitude of the orbital circles such as cannot be found in any other way. (DR 10^a)

We have seen that this is true. It is very difficult to alter the hypotheses of Copernicanism and save the theory. The different hypotheses of the theory are constructively confirmed by different data-sets. If the data had been contrary to some Copernican hypothesis, so as to constitute a counterexample, it is very difficult to alter just that hypothesis and save the bulk of the theory. We are forced to abandon nearly all of Copernicanism if we must abandon some of it. The data that we actually have observed are much more tightly constrained by the Copernican hypotheses than they are by the Ptolemaic ones, which can often be modified to capture some proposed non-actual data.

In Copernicanism,
A) it is easy to find consonant sets of hypotheses, sets with multiple independent justifications from the data and
B) these consonant sets are highly integrated. The justifications for any of the hypotheses in the set frequently depend upon other elements.

By contrast, Ptolemy does much less well on all these scores. Features *A* and *B* can make a good claim to be the constructive version of the virtue that Copernicus called the harmony of the theory.

Well, so what? Why should the real world, the external world that is independent of humans, somehow preferentially render true an interconnected set of hypotheses as opposed to a disconnected set? Why should it somehow be so as to make a "sensitive" theory true, when that kind of theory is difficult to reconcile with different data, instead of an insensitive one? Isn't interdependence a pragmatic, and not an empirical or semantic virtue?

I will argue in Chapter 11 that, construed in the way that constructivism requires, the kind of harmony that Copernicus cited is *not* a pragmatic virtue. Contemplating the falsehood of a theory with this feature, we are driven towards the view that observation reveals nothing about the world, so that natural science (and our common practices) become impossible.

Copernicanism is a "sensitive" theory because of A) and B). Take Glymour's three hypotheses as an example. They are consonant, each has one way of justifying it that depends on two of the others, and these justifications justify the numbers as one among a wide selection of alternatives. So if you alter any one piece of evidence, that usually determines a different value for one hypothesis immediately. Because this hypothesis is altered, it no longer works to determine the same value for other hypotheses in the set, and because these hypotheses have multiple means of justification from other pieces of data, these additional data prevent you from saving the hypothesis.

So one can identify a set of consonant hypotheses, by virtue of A), so that any alteration in any one of them would result in different observations than we have in fact observed. For every H in the consonant set, although $\{E\}$ only fallibly justifies H using $\{A\}$, there is some other justification for H which will work even if $\{E\}$ and some of the A are false. Put differently, if any H is false, a number of things we have observed are affected, not just those in a single justification of H. Other justifications for H must have gone wrong too, in a way we haven't detected.

In addition, in virtue of feature B, justifications of other hypotheses, H', that use H as an auxiliary, must also be wrong. For when we change the value of an auxiliary (H), we get a different result concerning the target hypothesis (H'). A formerly confirmed hypothesis, H', is now refuted. So not only the evidence confirming H must confirm something false, the reasoning that confirmed H' must somehow have gone astray too. If we have feature B, then there will be many hypotheses to play the role of H'.

In short, if we have a theory with features A and B – a harmonious theory – then supposing that one hypothesis is really false in spite of the evidence undermines a great deal. Multiple sets of evidence that agree in justifying H must *appear* to do so, but in fact fail. And multiple other pieces of reasoning which use H, when in fact it is false, must somehow end up all agreeing with each other about what is true in spite of the fact that they depend upon a false auxiliary.

All that is incredible. As the Reverend Charles Kingsley put it, in quite another context: "I cannot . . . believe that God has written on the rocks one enormous and superfluous lie for all mankind" (in a letter to Philip Gosse, quoted in Gosse 1890, 281). Kingsley was talking about some of the evidence that Darwin later used to justify evolution, but the point applies just as well to any set of harmonious hypotheses. The agreement of the evidence, from independent justifications, where the agreement cannot be attributed to cherry-picking, deceit, or other shenanigans on the part of human beings, leads us to the conclusion that we cannot trust reasoning from the (apparent) outcomes of observation to justify anything. It's no wonder that those with religious faith rebel against such a suggestion, and the more secular of us ought to too.

Copernicus' view, as he originally stated it, was wrong about a great many hypotheses. The planets do not execute epicycles, nothing in the heavens travels in circles, farther objects do not always appear to travel more slowly, the analysis of retrograde motion requires mathematical techniques he did not possess, and on and on. So it looks to be very weak to argue that he illustrates something that is a central feature of the conviction that evidence ought to carry today.

But I think that, upon reflection, the method of Copernicus is not as dated as his theory is. Far from being a passing phase of astronomical theory, this feature of astronomy – providing independent justifications for hypotheses – strengthened until the present day. The poor methods of observation that were available to Copernicus gave way to much superior methods used by Brahe, and eventually to the telescope. Much that Copernicus believed was refuted by these later developments. But the theories that replaced him, by Kepler and eventually Newton, followed him in requiring multiple methods to confirm consonant hypotheses.

So the advantage of Copernicanism is not that its harmonious nature is aesthetically attractive, a notion of which Kuhn was justly dubious (1957, 180). There is more than just an aesthetic preference for Copernicanism. Copernicanism raises the price of the falsehood of its hypotheses. If one of them is false, then we are massively deceived about the heavens, in a way that covers virtually all the phenomena we have observed. If this kind of harmoniousness doesn't count as an indication that belief is justified, then the outcomes of observation cannot justify belief in anything. The opponents of empiricism think it goes too far in depending *only* on observation for justification, but even they do not claim that one cannot use observation *at all*. That would make both natural science, as well as the common practice of living, impossible.

8 The Theory-Ladenness of Observation

The theory-ladenness of observation is the view that for an observation to have some outcome requires the truth of some additional hypotheses. To observe that something is an ammeter requires that there be such a thing as electric currents. To hear a hermit thrush requires that different species have characteristic songs. For it to be true that we have observed something, certain additional hypotheses, *backing-hypotheses* for the observation, must be true.

Hardly anything is universally agreed in the philosophy of science, but the idea that observations are theory-laden must come close. The idea that there was a theory-neutral observation language to which we could compare the edicts of scientific theories disappeared in the 1960s, debunked most influentially by Norwood Hanson, Grover Maxwell and Wilfred Sellars (Hanson 1958, 4–19; Maxwell 1962; Sellars 1963, 127–196). For a judgment that an observation occurred to be true, other hypotheses must be true too, and the same holds for an observation being probable, or being judged probable by a human. Constructivism wholeheartedly agrees.

The theory-ladenness of observation shows, according to the Quine-Duhem point of view, that constructivism must be wrong in ending justifications with the outcomes of observation. For the theory-ladenness of observation shows that we *cannot* appeal to the outcomes that observation has had without some additional justification for another hypothesis. For an observation to have some outcome requires the truth of some additional hypotheses. Whence, if an observation threatens our favorite hypothesis, we can alter the backing hypotheses so that the observation is no longer made, or is characterized in such a way as to avoid refuting the hypothesis. With appropriate changes, the same thing goes for confirming evidence for a hypothesis we dislike. So the Quine-Duhem hypothesis wins again.

After a brief review of this criticism of constructivism, the chapter goes on to reply to it. The substance of the reply is that theory-ladenness cannot prevent an outcome of observation occurring that refutes hypotheses of a theory, instead of some different outcome that does not. Furthermore, this is evident to both partisans and opponents of the theory because we do not have to *believe* the backing hypotheses in order to detect which of two outcomes happened. So theory-ladenness does nothing to prevent partisans pointing out to their opponents that an outcome of observation refutes (or confirms) a hypothesis, even if those opponents do not hold the backing-hypotheses for that observation. Enough of these inconvenient outcomes, with no compensating examples on the other side, makes it evident to the opponents that they face a problem. The ordinary, unreflective way of putting this problem is that the evidence is coming out in a way that refutes their view.

The opponents might *go on* to attempt to try to evade the problem by fault-tracing to some backing-hypothesis favored by the partisans. But they only do so because they've already recognized that the observations did not have the outcomes that favor their views. And fault-tracing isn't guaranteed to save the hypotheses which the opponents prefer; it depends on the cooperation of subsequent outcomes of observations. If these fail to cooperate, the upshot will be a refutation by theory-laden observations. The result is particularly telling if the partisans can produce independent evidence that favors their own backing-hypotheses, and the opponents cannot.

This gives a much better view of the way science is practiced than one based on Quine-Duhem. It's confirmed by the observations about how scientists go about constructing detection instruments. It also predicts that some methods for confirming a hypothesis will be legitimately rejected, because the justification offered for the backing-hypotheses results in a cycle. That will be the topic of the next chapter.

8.1 The criticism from the theory-ladenness of observation

Donald Gillies put the point succinctly:

> . . .that all observation . . . is theory-laden reinforces the holistic thesis [of Duhem]. [Consider a hypothesis] H_1, which could not be refuted by observation when taken in isolation, but only when taken as part of a conjunction of a group, G, of hypotheses, where $G = \{H_2, H_3, \ldots H_n\}$ say. Now suppose G is refuted by an observation statement, O. This statement, O, is established by the interpretation of sensations in terms of a further group of hypotheses, G', where $G' = \{K_1, K_2, \ldots K_s\}$, say. Thus, to test H_1 we need not only the hypotheses $H_2 \ldots H_n$, but also the hypotheses $K_1, \ldots K_s$. [A] scientist has, in addition to the option of changing one or more of the hypotheses in G, the option of querying one of the assumptions in G', in such a way that O is rejected. (Gilies 1993, 137)

But now, clearly, we are faced with the same regress that we faced with the endless introduction of auxiliary hypotheses. Nothing, apparently, can be known to be justified independently of anything else.

A constructive tree is supposed to be one complete justification for its target hypothesis – a real justification as opposed to a relative one, as Edidin put it. But if the outcomes of observation upon which the justification depend themselves depend upon their backing hypotheses, in what sense can this be a complete justification? Why is it any use giving evidence for the auxiliaries if we are not also given the evidence for the backing hypotheses? Why does a constructive tree answer any question that is worth asking and answering, given that it is relative, not to the auxiliaries, but to the backing hypotheses? This is what the Quine-Duhem objection from the theory-ladenness of observations boils down to.

8.2 Theory-ladenness must allow for unexpected outcomes

Israel Scheffler gave the clearest justification for this. He began by formulating a problem about observation. The ability for us humans to make a particular observation in natural science cannot be independent of what beliefs we have. A change in our own beliefs, possibly unprompted by additional observations, could lead us to withdraw some observation (Scheffler 1982, 36). The problem is that this conclusion seems to destroy completely the ability of observation to act as an independent check upon which hypotheses we believe. Instead of observations constraining what hypotheses we ought to believe, the latter controls the former. "Observation contaminated by thought yields circular tests; observation uncontaminated by thought yields no tests at all" (Scheffler 1982, 14).

In reply to this, Scheffler draws upon a distinction:

> This distinction may be put in somewhat different ways [. . .] We may express it, for example, as a distinction between concepts on the one hand and propositions on the other, between general terms or predicates on the one hand and statements on the other, between a vocabulary [. . .] and a body of assertions [. . .] between categories or classes [. . .] and expectations or hypotheses as to category membership[.] (1982, 36)

Scheffler goes on to draw the distinction on which constructivism relies to reply to Quine-Duhem. We have the ability to detect the difference between the presence and absence of events that occasion an observation independently of various different ways we might characterize what it is we observed. Even when we do not believe the backing hypotheses for making an observation, we can still tell the difference between that observation being made or not *if* the backing hypothesis is true. We can do this, even though there is no such thing as making an observation without any backing-hypotheses.

Even the Quine-Duhem perspective on observation allows this. Look at Duhem's original argument:

> The only thing the experiment teaches us is that among the propositions used to predict the phenomenon, and to establish whether it would be produced, there is at least one error; but where this error lies is just what it does not tell us. (1914 [1996], 185; see also 187 and 188 for similar remarks).

Harold Brown pointed out that Duhem's maneuvers would never be required if theory trumped observation, because we'd never have anomalous observations

(1993, 558). Adolf Grünbaum and Israel Scheffler made the same point earlier; observations made according to a theory can still present problems for it (Grünbaum 1960, 75, 82; Scheffler 1982, 39). Since we know that there is at least one error as a result of observation, making observations must have an influence upon what we believe.

Kuhn toyed with the idea that the expectations that came along with practicing a paradigm so strongly affected observation that it prevented scientists from recognizing refuting data (1962 [1996], 62–64, 112–113). Just observing that the oscillations of a piece of iron carrying a mirror involves an interpretation just as much as measuring electrical resistance does (Kuhn 1962 [1996], 125–126). The hypothesis that the impedance of a coil is 50 ohms is just as much a hypothesis of a scientific theory as the hypothesis that the piece of iron is oscillating with a certain amplitude at 50 hertz. If either hypothesis is true, certain other hypotheses must be true too, and that sets off cycles or a regress.

The development of equipment for observation according to a paradigm led, Kuhn argued, to both a restriction of vision and a considerable resistance to paradigm change (1962 [1996], 64). Experiments with anomalous cards, such as a red spade, suggested that with a brief exposure, subjects falsely categorized it as one from the ordinary decks of cards. Uranus was observed on several occasions before Herschel identified it as a planet, but was taken to be a star (Kuhn 1962 [1996], 115). Hetherington (1983) provides other examples; we "observed" that the sun had a strong magnetic field when we expected it to have one, and the Mount Wilson observatory "observed" the redshift for Sirius B that Eddington predicted and asked them to look for.

Nobody denies that instances like this do occur, but they cannot occur so often, or so incorrigibly, that we are incapable of perceiving things that refute our theories. The errors that Hetherington cites were quickly corrected. Kuhn himself noted that, although electrostatic repulsion was not noted with early instruments, improvements produced unexpected observations without seeking them (1962 [1996], 14–15, 35). Alan Chalmers provided examples of experiments by Galileo, Faraday, Hertz, and Perrin which were all decisive in their results despite the contested theories of the day (1976 [1999], 23, 195–196, 33, 204).

8.3 We can distinguish between outcomes without justifying their necessary conditions

We can distinguish between successful and refuting outcomes even if we disbelieve, or are agnostic about, the necessary conditions for its being true that that outcome occurred.[16] As a slogan, one could say that we can detect without believing. When we detect without believing, we do not endorse one description of what we detected. That is, we do not endorse the description of the outcome that requires that some given backing hypotheses be true, but rather some different description, with different backing-hypotheses. But we can still tell the difference between the detection and the non-detection.

There are a number of considerations that make this plausible.

First of all, we know of examples. We can all tell when Joseph Priestly would have held that a match was emitting phlogiston and when he would have held that it wasn't. Yet obviously we do not endorse any observation as an emission of phlogiston. We know when Ptolemy would have said that the sun was moving through Taurus while the earth was stationary (Scheffler 1982, 35). Nobody now believes that electricity or heat are fluids, but that doesn't prevent us from identifying the observations that could be cited as justifications for these hypotheses, and which those theories would have underwritten as observations.

In all these cases, we can tell whether or not some outcome of observation happened, but we disbelieve, and accept no justification for, the backing hypotheses.

A second reason we know whether or not an outcome occurred, even if we disbelieve the backing hypotheses used to express it, depends upon fault-tracing. We often discover that an observation was in error, but there are two different *kinds* of error we can find.

We can, first of all, find that the observer did indeed detect an outcome that actually occurred, but used backing hypotheses for it that are false. On 9th January 1493, Columbus "observed" mermaids in the new world. Since they do not exist, the observation must have been in error. He observed them near what is now the Dominican Republic, and described them as less beautiful than is usually supposed. So he probably observed manatees. These were present in the region at the time, and fit the description well. So the observation is of a genuine event, the outcome occurred as opposed to not occurring, but Columbus misdescribed it because he used false backing-hypotheses.

By contrast, there are cases in which the backing-hypotheses are all in order, and the observers believe them, but for some reason we decide that the outcome cannot have occurred. The Israelites did not, many now believe, observe the moon to stand still over the valley of Aijalon. But we do not think they're wrong about the backing-hypotheses for such an observation. The outcome simply didn't happen. On 26 March 1859, Edmond Lescarbault "observed"

---

[16] Here as elsewhere I take it that extreme subjective probabilities, 0 or 1, are ruled out for the backing-hypotheses.

the transit of the planet Vulcan across the face of the sun, inside the orbit of Mercury. Vulcan had at that time been proposed to explain the anomalies in the orbit of Mercury. But there is no such planet. Lescarbault understood his telescope, and knew what it was to make such an observation, but cannot have actually witnessed that outcome.

Consider the former kind of case, where the observer used the wrong backing hypotheses, but did genuinely detect that one outcome occurred. We learn something from their testimony. We know that they did indeed detect that one outcome occurred and another did not . But they mis-described it. The observation was "laden" with the wrong theory. So they can competently distinguish that the outcome as we describe it occurred, even though they do not believe the backing-hypotheses for that outcome. Those who do not share our views can distinguish between the (correctly described) outcome happening and not happening, even though they do not believe the backing-hypotheses that would have allowed them to describe it correctly.

To see the subtlety and range which scientists have to identify the same objects, properties, and kinds of events in spite of completely different backing-hypotheses, consider an elegant example from Philip Kitcher, again dealing with the phlogiston theory.

Joseph Priestly observed, and breathed, dephlogisticated air in his laboratory (Kitcher 1978, 533). He heated the calx (that is, ore) of mercury (mercury (II) oxide), and collected the evolved air (gas). The ore turned into mercury. Because (according to Priestly) the metal contained phlogiston, he concluded that this air must contain very little phlogiston. Burning (combustion) was the emission of phlogiston. Since the phlogiston has already been removed from this air, he speculated that it would readily absorb phlogiston, and hence support burning better than ordinary air. He found this was true; burning in the dephlogisticated air (oxygen-enriched gas) was brighter and mice thrived in it. He breathed some of it, and "fancied that my breast felt peculiarly light and easy for some time afterwards" (Kitcher 1978, 533).

Almost everything in this story is observable; the calx, the dephlogisticated air, the burning and so on. It is easy to see how we would describe these observations, and the fact that Priestley frames them in ways that depend upon backing hypotheses that we do not believe is no impediment. Like Lavoisier, we can use Priestley's account to repeat the experiment. Warming mercury (II) oxide decomposes it into mercury and oxygen, and the latter supports combustion better than air does. Here, we fault-trace, not to the outcomes, but to the backing hypotheses for them. But still, we are able to identify which outcomes occurred, and which substances were observed.

A third reason for thinking that we can detect differences in outcomes even when we do not agree about backing hypotheses is that distinguishing between outcomes is an ability we possess, not a hypothesis we justify. We share this ability with non-human animals, and with other humans who do not possess the use of language. Other animals can certainly detect whether or not a predator is present, and some give every non-linguistic sign of being able to recognize individuals. Even people who are capable of more sophisticated judgment might be capable of simply arriving at an unreflective detection in addition. You can tell whether or not there's a pig in front of you. While this ability *might* require all sorts of beliefs in backing hypotheses, that hypothesis is not inevitable. It is perfectly reasonable to suspect that this is something we can simply do, as non-human animals can. (This reassuringly prosaic example is from J. L. Austin 1962, 114–117.) And there are variations in this ability that are not variations in commitment to backing hypotheses. Color-blind people and tone-deaf people have no problem understanding or believing backing hypotheses. They just cannot see or hear certain distinctions between outcomes that others can.

Fourth, one is sometimes forced to allow that, inconveniently, one's opponents are at least right about how things came out, even if they describe it the wrong way, and draw the wrong conclusions. Integrity demands that one acknowledge those occasions when one's opponents did an experiment and took a risk, and that this paid off so far as they were concerned. We can take bets on outcomes with even the most benighted ignoramus pedaling anti-scientific rubbish. Sometimes, both parties realize, one side lost the bet.

These conclusions about outcomes that favor a theory even though we do not agree with the backing hypotheses for it are evidently ones which we can and do draw; we spontaneously agree with each other about them. Moreover, we succeed in arranging our social interactions this way, as in the case of bets upon whether or not an experiment will have one outcome as opposed to another. It follows that we recognize something in common, when we interact with people who disagree with us about backing hypotheses. As I would put it, we both recognize that one outcome rather than another has occurred, but we endorse different hypotheses concerning what it was. That is, we describe an event we both recognize in different ways. So the same result follows; we do not have to share backing hypotheses in order to be able to recognize the difference between different outcomes.

8.4 Why constructive trees justify without a justification for backing-hypotheses

With this in hand, we can reply to the objection from Quine-Duhem. When one outcome of observation occurs rather than another, there is, first, an event in the external world of some kind. A balance balances, or a mouse survives in a sample of gas for a long time, or an interference pattern fails to shift. Other examples are more complicated; there are more dark moths in one area rather than another, or a microwave horn gives the same reading in any direction of the sky, or a mineral deposit is found both on the west coast of Africa and the east coast of South America. Put in mundane terms, observations require that something happen or obtain. The difference between it happening or obtaining and it not happening is what I call the difference between one outcome occurring and some other outcome occurring.

Some physical process caused by the outcome then affects our bodies, most obviously through our sense organs, although any effect will do. (One could observe by getting a tan, or by feeling light-headed.) As I have said, constructivism treats the human body as a kind of measuring instrument (van Fraassen 1980, 17). This is the second part of the process of observing; the outcome affects an observer.

As a result, we either come to some conclusion, or recognize that some conclusion is warranted if certain backing hypotheses are true. We see that the outcome of observation occurred as opposed to not doing so. The conclusion which an observer draws or recognizes ought to be drawn given certain backing-hypotheses is what I will call the *hypothesis of observation*. Understanding that some hypothesis of observation is warranted if the backing-hypotheses for it are true is the third part of making an observation. We cite the hypothesis of observation to others in communicating an empirical justification to them. The psychological processes that leads to our judgment that the hypothesis of observation is true is desperately obscure, but that doesn't make it doubtful that we do draw that kind of conclusion when we are affected in certain ways.

This is all very vague, but that is part of the point. It isn't supposed to be detailed, but rather a minimum characterization of what observations are. I hope it's vague enough to be uncontestable. It fits what theories of empirical confirmation say. It is standard in Bayesian analyses (for example) to see observation as a non-rational change in the probability of something – a hypothesis of observation or an observation sentence – followed by a rational process of Bayesian updating (Strevens 2012, 24).

Every hypothesis of observation is part of some scientific theory, and we are not justified in believing any hypothesis of observation unless other hypotheses

8.4 Why constructive trees justify without a justification for backing-hypotheses

are true, what I've called the backing-hypotheses. Given that every hypothesis of observation depends upon some backing-hypotheses, why is it legitimate to present constructive trees that end by citing them? Why should constructivism not also have to defend the backing-hypotheses in addition?

Because distinguishing which outcome occurred is something we can do independently of how confident we are in the backing-hypotheses. Affirming the hypothesis of observation conveys the information that the outcome of observation occurred rather than not occurring. And this change in probability does *not* require that the observer have any particular probability – high or low – in the backing-hypotheses. Even someone who strongly doubts the backing-hypotheses receives this information and is capable of using it to justify additional hypotheses. What happens when we discriminate that one outcome rather than another has occurred or is so is independent of our justification (if any) for the backing-hypotheses.

I'll work through an example of this answer in action. Thomas Kuhn gave a variety of reactions to the evidence the telescope provided in favor of Copernicanism (1957, 226). By looking through it, someone who believed it was a dependable instrument for better observing distant objects (D) would believe that Venus had phases (V). Not all observers did believe the telescope was dependable, as Kuhn pointed out. D is a backing hypothesis here. (So D is something like "We observe Venus when we look through a telescope pointed at it", or "The telescope is a **D**ependable way to observe distant objects".) If Bayesians assign D a prior probability of zero it is impossible to learn things from Bayesian conditionalization by using it, but the same is true for any other hypothesis. This is part of the Bayesian problem of zero priors, and is why, to avoid making Bayesianism into a straw man, priors are taken to be non-zero.

Let B be someone who follows Ptolemy and who has no evidence for or against the idea that the telescope is dependable. So B is a fanciful version of Bellarmine. So long as he does not set $Pr(D)$ equal to zero, he too will realize that $Pr_{new}(V) > Pr_{old}(V)$.

Richard Jeffrey gave an analysis of another situation in which observation raises probabilities, but not to unity (1965 [1983], 165). We often observe in candlelight or under other uncertain conditions. So we need a new way of calculating how the probability of hypotheses should change when we make an uncertain observation. This is *Jeffrey conditionalization*:

$$Pr_{new}(H) := Pr_{old}(H|E).Pr_{new}(E) + Pr_{old}(H|\neg E).Pr_{new}(\neg E).$$

As Jon Earman points out, this will be correct if and only if the conditional probabilities of the hypothesis given the evidence do not change upon observing (1992, 34–35).

Jeffrey conditionalization concerns the use of an observation made under uncertain circumstances. Here we have uncertain circumstances, because we are not sure whether a necessary condition for observing E is satisfied. I propose, therefore, to take the rise in probability of V given uncertainty about the telescope to function like observation under dim light, or other obscuring circumstances. This is quite standard in Bayesianism; Michael Strevens, again, uses Jeffrey conditionalization in this context (2012, 24). Jeffrey conditionalization is a way for us to go on to reason about H in spite of this uncertainty. So, for constructivism, H concerns the justification of some subsequent target hypothesis such as Venus orbiting the sun, by the observation made under uncertain conditions, namely that Venus has phases.

Under this simple analysis, then, both the Copernican and the follower of Ptolemy will raise the probability that Venus has phases, and by Jeffrey conditionalization raise the probability that Venus orbits the sun. The rise in probability need only be slight, but it's a start. We can use outcomes of observation to confirm, even if we do not go on to cite evidence for or against backing-hypotheses.

8.5 Replies to objections

In reacting to all this, an opponent of heliocentrism, B, might indeed decide that he hates heliocentrism so much that the telescope must be even *less* probably reliable than he formerly thought. That is, suppose he looks through the telescope and sees (apparently) a body that is illuminated from one side. He realizes that this means (according to his prior probability distribution) that he will have to raise $Pr(V)$ slightly. Horrified, he decides that $Pr(D)$ and $Pr(V|D)$ must be even lower than he'd thought, and so consequently he can protect $Pr(V)$. (So, as Earman pointed out, Jeffrey conditionalization doesn't work because the conditional probabilities have been altered.)

In such a situation, B is clearly observing *something*, although it cannot be V, or he'd be observing the phases of Venus. So we will call the something that B says he observes X. X looks suspiciously like sense-data. Under the foundationalist way of analyzing things, we could say that anyone who looks through the telescope sees sense data, which then raise or lower the probability of V using D as background knowledge. We would then get the usual Bayesian analysis, with the sense data being the evidence, V the target hypothesis and D the auxiliary. To avoid making B into a straw man, we will suppose B would say X was a publicly observable pattern of light from the eyepiece (or something) so he's not embroiled in foundationalism. What B is doing here is reasoning that since the telescope suggests that something is true which he knows must be false, this makes it unlikely that the telescope is a dependable means of observation. That seems reasonable.

The Bayesian analysis of the situation was given by Jon Dorling (1979). Dorling was interested in the case where a hypothesis at the core of a theory and which is strongly believed (here, $\neg V$), when conjoined to an auxiliary (the dependability of the telescope, D), predicts and observation, $\neg X$, which gets falsified. In such a case, Dorling points out, Bayesianism blames the auxiliary, not the target hypothesis. Dorling's conditions are satisfied in this example.[17] So does Quine-Duhem come out on top? Must we justify the backing-hypotheses for an observation in order to have a justification?

Not at all. According to Dorling's analysis, while a Bayesian B can indeed focus *most* of the blame on D as opposed to V, he does have to raise the probability of V. This is just a consequence of being a Bayesianism and getting an unwelcome outcome. So long as the new probabilities have to be the old conditional on observation, some blame, although perhaps little, must attach to your favorite hypothesis. But that is all constructivism wants or needs. More and more negative evidence that is independent of D, even if that evidence has doubtful backing hypotheses, can refute a hypothesis in spite of partisan preference for it. Also, of course, D may get justification from sources independent of astronomy.

Perhaps the idea is that B can adopt a completely new probability distribution, one that keeps the probability of V where it was, or even lowers it. The most serious difficulty with the proposal is that if this is permitted, we permit far too much to count as the practice of natural science. We can save hypotheses in the face of evidence, to be sure, but only by allowing anyone to believe almost anything at any time. There are some limits, because the new probability distribution has to be coherent. But you cannot say that priors are washed out by the evidence, because a whole new set of priors can appear any time anyone doesn't like the way things are going. If this is the only way to save Quine-Duhem, then that hypothesis becomes irrelevant to our understanding of what we actually do in natural science.

Arguing like this, though, really misses the central point that constructivism wants to make here. So I'll just suppose that the Quinean can somehow figure out a way to make $Pr_{new}(V) \leq Pr_{old}(V)$ without abandoning Bayesianism. It doesn't follow from this that B cannot comprehend what Galileo is driving at when he points out that V rather than $\neg V$ happened. Even if B doesn't alter his

---

[17] The reader may check that the following priors hold: $Pr(\neg X|D \wedge \neg V) \approx 1$; we observe X, which is anomalous for B; $Pr(D)=Pr(D|V)$; $Pr(\neg V) \gg Pr(D)$; $Pr(\neg X|V \wedge D) \ll Pr(\neg X|V \wedge \neg T)$ and $Pr(\neg X|V \wedge T) \ll 1$. Dorling shows that the posterior of D drops a lot, while $\neg V$ only drops a little.

posterior probability of *V*, because he decreases his posterior for *D*, he still knows that it is true that the one outcome, rather than the other, obtained. And he knows that Galileo's beginning a justification with *V* signals that fact, which he is bound to acknowledge, whatever either of them thinks about how that ought to affect *D* or whether either is compelled to raise *Pr(V)*.

That is sufficient to show that Quine-Duhem is false, because if it happens enough, then B will have to allow that the evidence came out in a way that justifies Gallileo's hypotheses over Ptolemy's. Even if B holds on to the same *Pr(V)* in spite of a long record of refutation, this is not holding on to any hypothesis no matter how the data come out – it is holding onto a hypothesis *in spite of* the way the data come out. Quine-Duhem should affect the philosophy of science because it appeared to present a way of coping with new data *in accordance with* some way of using data to confirm hypotheses that we observe in natural science. If it's just *ignoring* data, then we should ignore *it*, at least insofar as we are trying to understand how natural science delivers knowledge.[18]

According to constructivism, whether or not the backing hypotheses are indeed true is a matter of *subsequent* fault-tracing. That is, it is a matter of whether Galileo or B can muster other observations to render *D* more or less probable independently of *V*. Given that *D* is true, and the telescope is dependable, the outcome of observing might be *V* or ¬*V* prior to observation, and the act of observing rules one of these out. That is all that constructivism requires in order to increase the probability of some subsequent hypothesis, for example, that Venus orbits the sun. All this is glaringly obvious to B, or he'd have no motivation to *attack* that backing-hypothesis.

Let's move on to a different objection. Some people might say that if two people have different backing-hypotheses for observation, then they cannot observe the same outcome at all. Outcomes are supposed to be events or processes in the world, outside a human body (if the observation is not about the human body itself). But there are no such outcomes. There is no one thing, or kind of thing, that different scientific traditions can both observe. There is no distinction, such as that between one outcome or another occurring, that they can both recognize. They respond to different worlds, as Kuhn once said (1996 [1962], 111).

I have two replies. The first is from Israel Scheffler (1982, 40–41). Even if two groups of scientists differ in the backing hypotheses that they use for some hypotheses of observation, it does not follow that they cannot share

---

18 At one point in the debate, Quine did agree that his thesis was trivial (Harding 1977, 132). But he never maintained that it was false of the justifications offered in natural science. Had he done so, his thesis would have been irrelevant to anything which any version of empiricism maintains.

8.5 Replies to objections — 157

other backing hypotheses that will allow them commonly to identify which outcome of observation occurred. To give a simple example once more, Copernicus will not accept the backing-hypotheses behind Ptolemy's observation that the sun is rising, and so will disagree with the latter's assertion that it is. For the same reason, Ptolemy cannot agree that the earth is rotating into the sunlight. Both, though, can agree that a hot, bright yellow, apparently circular object is now visible in the sky, when it was not before. This agreement gives some sense to the idea that they can observe the same outcome of observation, for both can see something that prompted the other to assert their hypothesis of observation.

I would, secondly, point out that whatever it means to say that different groups of scientists live in different worlds, they can still do something which looks a lot like communicating and repeating each other's experiments. As a result, they can use each other's work to acquire justifications.

In the example above, for instance, Priestly communicated his experiment to Lavoisier, who used the same method to generate the same gas (or at least, that is what one would ordinarily say about what happened). The objection denies that this counts as seeing the same things. But we can identify things that Priestly saw ('calx', for example) which we can see plays an analogous role in his justification to things Lavoisier used in his justification ('mercuric oxide'). (Kitcher argues that these words refer to the same kind of thing, even if they have different senses (1978).) As long as we can do that, it hardly matters whether we say these things are the same or different. Lavoisier can still identify whether or not an outcome happens, and Priestly still led him to that identification. We can distinguish Lavoisier from someone who (as we'd put it) misunderstood Priestley and did some utterly different experiment. We can distinguish between Lavoisier getting the same results as Priestly (as we'd put it), and his getting different ones. So we still know when to use the expression 'repeated the experiment and got the same outcome' and when not. We have what we need to get on with the practice of scientific justification.

One final objection. There will be some, in the light of all this, who want to say that outcomes of observation are nothing more than disguised foundations, and that I'm a secret foundationalist. I've repeatedly had to resort to expressions such as 'distinguishing between a thing happening and not' and 'understanding that other scientists have observed something without agreeing to the description that they give of it' to indicate what I mean. These suggest that there is one thing, a something, that happens whenever two scientists understand one another as faced with one outcome of observation as opposed to another. So the idea is that I have a secret ontology of events, or whatever outcomes of observation are, to which justifications are reduced.

I have to adopt some language or other to communicate what I'm driving at. I've chosen words like 'which outcome happens' because they appear to me most appropriate to the task. And my position is immune to the criticisms of foundationalism in the literature. Outcomes of observation do not deliver incorrigible knowledge, they do not depend upon sense data, they are apt to real observations made by real scientists, and they have backing-hypotheses as preconditions. We must all allow for the fact that we have the abilities to engage in science, and that it's a public practice, and it's that fact that has caused me to write as I do.

So the objection alleges that I must be committed to the view that there is *something* in common that both Priestly and Lavoisier observe when the one observes phlogiston emission and the other observes combustion. For this something must be the foundation on which the objection relies.

But I do not see that anything we observe about the practice of science, or the behavior of Priestley and Lavoisier, confirms the hypothesis that:

There is something that they both observe when Priestley observes phlogiston and Lavoisier observes combustion *in any sense that will support the objection*.

All we know is that (as we'd put it) Lavoisier was pretty much right about combustion, and Priestley's theory was clearly refuted; that Priestley did an experiment, wrote things down, and Lavoisier repeated the experiment; that when combustion happens, both men can detect it, as is evidenced by their behavior, but make different noises and smear different patterns of ink, as is also evidenced by their behavior. Because we all share this ability to detect, we know now that Lavoisier was pretty much right, and Priestley very much wrong, in the noises and ink department. So if we describe things in the natural way, the sense in which there is one thing that Lavoisier and Priestley both observed, is that they both observed combustion.

None of that supports the objection, because according to the view in the last paragraph, Priestley *didn't* observe phlogiston, because there's no such thing. According to the objection, I have to hold that there is something they both observed (which I do hold – combustion) *and* that somehow it's *a-thing-outside-there-in-the-world* which is at the same time *one* thing, but also makes *both* Priestley right about it, *and* Lavoisier right about it.

I do not hold this. Instead I hold that Priestley didn't see phlogiston, but did see an instance of combustion. I do not see that there's any evidence in the behavior of the two, or our ability to communicate, take bets, and engage in science, that confirms the existence of this something in which the objection says I must believe.

Even when we have no good idea who is right, or what the right answer is, it doesn't seem to me that my view has the consequence that there is some atheoretical thing-in-itself to account for our shared abilities. Suppose we do not know what we are observing or disagree about it. But the 'it' about which we disagree isn't some atheoretical thing in itself. When groups of scientists are capable of being puzzled about some phenomenon which they all observe, then in order for them all to observe it, there must be enough commonality between them to agree sometimes about whether the phenomenon is happening or not . It seems to me that everyone has to agree to this, otherwise there would be no way to understand what's happening as a common phenomenon puzzling both. But just like everyone else, constructivism can hold that when two communities are capable of coming into agreement about whether or not something is happening, when they disagree about exactly what it is, they do not both access some atheoretically characterized event. They agree about some characterizations ("the hot bright event that is happening now, on this bench") and disagree about others ("the magnesium is emitting phlogiston"/"the magnesium is undergoing combustion").

9 Cycles of Observations

The backing hypotheses for observations are not a separate group of hypotheses. One hypothesis plays different roles in different justifications. Now as a target, now as a backing-hypothesis, and again as an auxiliary. We have already seen how backing-hypotheses get confirmed in the example of the ruler and the scale. Some outcomes of observation that we made without the aid of instruments justified using a balance to measure forces. The instruments we make can then be used to show that *other* methods of unaided sensation are reliable under some circumstances. So we can use measuring instruments to reassure ourselves that our unaided senses are reliable. So too, we can confirm hypotheses of the way in which our senses work, and integrate the functioning of our own senses into our theories of the way that the processes of the world work.

But clearly, according to constructivism, we ought to doubt a justification for a measurement to the degree that it is involved in a cycle. This chapter defends the role of these prohibitions on cycles of justification when it comes to backing-hypotheses.

9.1 Why constructivism gives a better account of observation

Constructivism gives a well-motivated account of our rejection of one kind of vacuous observations: those involved in a cycle. Quine-Duhem cannot duplicate that account. Under Quine-Duhem it is puzzling that we should even find these kinds of observations odd. The suspect kind of observations are those which require backing-hypotheses that produce a cycle of justification with the target hypothesis. To prepare the ground for making that point, we need to look at some general features about observation in science.

Constructivism says that a justification *can* be genuine without an additional justification for the backing hypotheses of the observations. But it allows that it's better if we *do* have such an additional justification. And in normal science, we almost invariably do.

Constructivism says that if we were to investigate, we would find that scientists know of experiments that justify the use of radio-telescopes, and particle accelerators, for making observations. We must, or we would not imbue their deliverances with the authority that we do.

Constructivism claims that these justifications are independent of the hypotheses these instruments are used to confirm. The constructive trees, and the hypotheses upon which they depend, cannot make a confirming outcome of observation inevitable, nor prohibit it. That is quite plausible though. People who actually work with these instruments do concern themselves with making sure that they're reliable, and testing their outcomes. Constructivism predicts that this requires a series of independent, constructively analyzable, experiments. (The point is not restricted to instruments used for observation. When we prepare a source of (for example) particles in a particular state, we will need to check repeatedly that they are in that state, in a way that doesn't prejudice the results in the experiments these particles analyze.)

Ian Hacking was the first to emphasize features of scientific practice concerning the experimental testing of instruments and other issues surrounding the making of observations. Scientists test instruments for observation in a variety of ways to determine their range of sensitivity, precisely what they are sensitive to, under what conditions they are sensitive to it, and how accurate they are. His *Representing and Intervening* (1983) contains, for example, an extensive, careful, and thorough treatment of a variety of different ways of testing detection-instruments. These cover, for example, the use of microscopes (186–209), investigations of observations of the sun's light and heat (176–80), and Faraday's observations on the interactions of electric currents and magnetism (210–11). Hacking's work inaugurated the emphasis on experimentation, instrumentation and laboratory practices that saw the observations they created as neither unproblematic givens nor dictated by adherence to a theory. In Deborah Mayo's words, ". . . experimental evidence need not be theory laden in any way that invalidates its various roles in grounding experimental arguments. [. . .] Some have especially stressed the independent grounding afforded by knowledge of instruments" (1996, 62).

It's now uncontroversial to point out the kind of subsidiary experimentation that sets up a preliminary context for some subsequent test of a target hypothesis. What is new is the claim that this kind of initial brush-clearing should be analyzed constructively. We have two categories of experiments. First, there are experiments to establish that some experimental technique for generating observations really is reliable. When the outcome of observation obtains, we can use the technique to conclude correctly that it obtains, and when it doesn't, we have a high probability of detecting that too. Then there's a target hypothesis and some experiment used to test it. So, for example, Peter Galison described the way CERN had one community of scientists building and testing a bubble-chamber prior to an overlapping group of scientists trying to use it to detect neutral currents (Galison 1987, 135–233).

What constructivism claims is that the experiments establishing observational reliability must be constructively analyzable. They must be constructively analyzable *in a way that is independent of the target experiment the observations are used for*. An observational technique is reliable when, both, if a specified outcome of observation occurs or is so, we know it occurs or is so (either certainly or very probably), and when that specified outcome is absent, we know it is absent. A set of backing-hypotheses for a kind of observation is some set that is sufficient for the reliability of some observational technique.

Return to what constructivism says about the independent justification of auxiliaries. We could not have a set of justifications for auxiliaries, the result of which was that the negation of the target hypothesis had a probability of zero. This was *hypothesis independence*. By the same token, we cannot have justifications for the reliability of observations in some justification that result in an agent believing that the target hypothesis must be true.

Justifications for auxiliaries were *observation independent* when accepting them could not result in the probability of the confirming outcome of observation being one. In the same way, justifications for the reliability of some observational techniques are independent of a target justification when accepting them allows a non-zero probability of a refuting outcome of observation for the target.

These conditions can be extended in a natural way. Hypothesis independence says that the background knowledge used to make some observation cannot guarantee the truth of the tested hypothesis. By extension, it cannot require its falsehood either. Observation independence says that a confirming outcome that some observation has cannot be inevitable for anyone who accepts the background knowledge for making that observation. By extension, neither can a refuting outcome be inevitable. The conditions are motivated by the idea of what an empirical test is. It's any circumstance in which an observer can be initially in doubt about whether a target hypothesis is true or not, and use the outcomes of observation rationally to alleviate that doubt.

If constructivism is correct, violations of observation and hypothesis independence ought to look wild as justifications, and non-violations should look more plausible. We have seen several examples of instruments for observation that have been shown to be reliable by constructive methods. Do violations of independence for the backing-hypotheses for observation look wild?

9.2 Examples of vicious background-dependence in observations

Alan Chalmers gave a very nice example of a viciously background-dependent kind of observation:

> I [once] taught physics at school level. I remember being troubled by an experiment I was obliged to have a senior class conduct. It involved measuring the deflection of a coil

suspended between the poles of a magnet as a function of the current passing through it. What troubled me was that I knew what was inside the ammeter being used to measure the current, namely, a coil suspended between the poles of a magnet. In this experiment the deflection would be proportional to the current readings whatever the relationship between current and deflection provided that both coils were governed by the same relation. (Chalmers 2003, 495).

Clearly, such an experiment could not confirm that the deflection was related linearly to current. From background knowledge antecedent to the experiment, it follows that no counter evidence will be observed even if the relationship is non-linear. Harold Brown gave a similar example concerning Special Relativity – its analysis of the Doppler effect prevents any observation refuting the maximum speed it attributes to a source (Brown 1993, 556). We saw, too, that Hasok Chang made a similar case against testing the linear expansion of a thermometric fluid by a certain experiment (Chang 2004, 59).

The examples violate *evidence independence*. That is, given the justifications we possess for the reliability of the methods of observation, the experiment cannot refute its target.

Chalmers, for example, knows that there is overwhelming support for three hypotheses:
1. All coils obey the same relationship between current in a magnetic field and torque on the coil.
2. There is a coil in a magnet in the ammeters.
3. The ammeters are calibrated so that the change in angle of deflection will indicate a proportional change in the current.

From these it follows that the probability of observing anything but a linear relationship in the experiment the students did is almost zero. Only 'almost' because something might still go wrong – an ammeter might be broken, or a student might be careless about measuring the deflection. But the experiment the students really performed wasn't the one they were told they were performing. It wasn't testing the hypothesis that there is a linear relationship between current and deflection. It was testing whether the ammeters worked or whether the students were competent.

Chang's example is a simple way to illustrate the same point. If a thermometer is calibrated by taking two fixed points and then dividing the interval between them into equal parts, then clearly, observing a fixed change in volume with respect to a fixed change in measured temperature does nothing to demonstrate that thermal expansion is linear. In such circumstances, what will be observed as a fixed change in temperature is simply a fixed change in the volume of a fluid. Some independent means of accessing temperature difference is needed to establish the claim that fluids expand linearly.

9.3 The literature on the topic

The solution I've suggested fits well with what others have said about observation. John Greenwood, for example, writes:

> . . .the observations that confirm our best current explanatory theories are informed by *quite different theories* which enable us to make the critical observations that establish our best current explanatory theories. The theory-informity of observation would only pose a threat to objectivity if observations are *necessarily* informed by the explanatory theory (or theories) which is (or are) the object of observational evaluation (Greenwood 1990, 560, italics original).

This is a version of the solution first argued in detail by Peter Kosso (1988, 1989, 1989a). Others have followed Kosso's general idea. There is a very nice treatment by Martin Carrier (1989). Ian Hacking mentions the solution briefly (1983, 183–185). So does Eliot Sober (1999). Robert Hudson (1994) dubs this *background independence*. It is a class of replies to the view that theory dependence entails a viciously circular test of any hypothesis or theory.

Kosso's version of the background independence reply is typical. Observations, he urges, only test a theory when they are not *nepotistic*, as he terms it (Kosso 1988, 464; 1989a, 136). If a means for securing an observation is to provide evidence for a hypothesis of a theory, then the target theory must be independent of the theories that are necessary to justify that the instrument detects what it's supposed to (Kosso 1988, 464).

This solution has the virtue of ruling out some clearly circular cases, and of permitting many clearly legitimate examples. It also has the advantage of clearly capturing the way scientists commonly use the term 'observation' and related words, as for example, when one says that the top quark was observed after it had been predicted (Kosso 1989a).

9.4 Chalmers' example; the constructivist analysis

Constructivism can make sense of Chalmers' example, and doing so provides an illustration of the constructivist view of observation.

Chalmers had a misgiving because the experiment presented a purported justification to the students which it did not really justify. The purported justification must have been something like this:

> Target hypothesis: the torque on a coil suspended in a magnetic field is proportional to the current in it.

Observations: We observed current $j1$, $j2$, $j3$ etc. and deflections $\theta_{coil}1$, $\theta_{coil}2$, $\theta_{coil}3$ etc.

Auxiliaries:
1. The coil is attached to a spring, and θ_{coil} is proportional to the torque on the spring.
2. The coil is suspended in a magnetic field.

Details of the justification: When you plot $j1$, $j2$, $j3$ etc. against $\theta_{coil}1$, $\theta_{coil}2$, $\theta_{coil}3$ etc. you get a straight line with an x intercept at zero. So, in this case j = constant. torque(coil). We generalize from this example to others.

We have no problem justifying these two auxiliaries independently of the hypothesis and the observed outcomes. Chalmers' misgiving concerned the backing hypotheses for the use of an instrument to measure the current, j. To get a full analysis of the background knowledge, we would also need to justify:

Auxiliary3: The deflection of the ammeter, $\theta_{ammeter}$, is proportional to the current in the circuit.

By simply looking at the faces of the ammeters, we can see that they are calibrated in such a way as to require that Auxiliary3 is true. Double the deflection leads to the conclusion that double the current is flowing in the circuit.

Chalmers' objection is that the students are not being made aware of the need for Auxiliary3, and that once one is aware of it, the purported justification for the target hypothesis cannot be made constructive. The design of the ammeter requires that the target hypothesis is true. As Chalmers observes, so long as the set-up in the ammeter and the apparatus obey the same law, any way of justifying Auxiliary3 results in:

$$\Pr(\neg \text{Targethypothesis}|\text{Auxiliary3}) = 0.$$

This result is explicitly prohibited by the definition of constructive Bayesian hypothesis independence from chapter two. Justifications cannot result in that equation according to the definition.

One can put the same point in terms of evidence independence. Suppose that Auxiliary3 is justified. What is the probability, then, that we will observe the outcomes in the justification above? What is the probability that the observed deflection of the coil will be proportional to the observed current? Well, so long as both set-ups obey the same rule, clearly unity. So regarding Auxiliary3 as justified entails that:

$$\Pr(\text{observations in purported justification}|\text{Auxililiary3}) = 1.$$

This, again, is just what is prohibited by the definition of constructive Bayesian evidence independence from chapter 2.

The set-up is not completely useless. It does show that the relationship between torque and current for two coils in a magnetic field must be of the same form, and it places some limits on that form. But that wasn't what the students were told that it justified. They were told that it showed this relationship must be linear. It shows no such thing, because it cannot be made into a real justification. It cannot be made constructive.

I think it is possible to justify Auxiliary3 independently of the target hypothesis. There are other ways of detecting current besides this design of ammeter. But of course, once one does this, the main experiment becomes largely pointless. Once we have justified the auxiliary, we have also justified the target hypothesis, since Auxiliary 3 turns out to be an instance of the target hypothesis.

I think that the ability of constructivism to make sense of examples that are, intuitively, not examples of genuine justification, is limited evidence for constructivism being necessary for real confirmation.

9.5 Conclusion

If Quine-Duhem were really the correct analysis of background knowledge, cycles of confirmation would be all but inevitable. The Quine-Duhem hypothesis alleges that in order to make observations, we require some theory. When we try to check that theory, we can do so only by making observations, and of course those will also require a theory. Since there is no end to this, cycles of confirmation would be all but inevitable when it comes to investigating the background for making observations.

It is then a considerable puzzle that so many tests that appear vacuous depend upon cycles of justification. There really ought to be no problem with such cycles, since they should be present, upon examination, even in those justifications that are beyond reproach.

Constructivism cannot allow cycles of confirmation. There must be some contribution to justification that derives from the fact that we detect one outcome and not another when we make an observation. We have seen that, if we ignore the justification for the backing-hypotheses for observation, this contribution is likely to be small. Still, it does exist.

We do not begin from nothing in doing real science; we are afloat in Neurath's boat, as Quine liked to say. All the same, I do not see why science could not be taught to children as if it does proceed in steps, beginning with a child's view of observation, and introducing the hypotheses of natural science only along with outcomes that justify them. Even the child's original hypotheses of observation might be justified or replaced this way. We cannot begin observations from nothing, but if the items from which we begin can be subsequently justified or modified, then we might eventually be able say we have a justification in the outcomes of observation for all that we believe about how to observe. Once again, we couldn't justify the whole thing at once, but might be able to justify any single part we desired.

10 Van Fraassen's Paradox

Constructivism, like every other version of empiricism, holds that observation is the only source of our knowledge of the physical Universe. In particular, it holds that any justification for some item of knowledge must begin with outcomes of observation.

Well then, why isn't it saying that the only thing we really know is *just those outcomes we detected*? It might do this, for example, by distinguishing between observable and unobservable objects, properties, and events. Inspired by van Fraassen (1980, 1989), it could point out that the faces of ammeters, febrile patients with black urine, and the earth, are all observable. By contrast, low energy electrons, the malaria parasite, and continental drift are not. Then we could believe that the ammeter was indicating a certain number, but not commit ourselves to the number of electrons moving in the circuit. A certain theory, one we accept, says that there's a current of 0.3 mA, carried by a certain number of electrons per second. We do not believe this part of the theory; we simply do not know whether or not it's true. We believe only what it says about those events we *can* observe.

There is an excellent argument for this restriction on belief, given by van Fraassen, and which I name for him. It is that by believing only what a theory says about the observables, and being agnostic about the rest, we expose ourselves to just as much risk of being found wrong, because only observation can show us to be wrong. At the same time, we are able actively to pursue science in the way a practicing scientist does. And at the same time as securing those advantages, we believe less, so we cannot be more likely to believe falsehoods. We appear to believe everything science can be right or wrong about, without believing all that some others do. This strongly suggests that the extra that these others believe isn't really justified at all. It is costless to drop this extra belief, so far as the information we are capable of actually getting is concerned.

The argument is independent of the details of the way in which the distinction between the observable and unobservable is drawn, so long as it can be coherently made out in a way that puts observations on only one side. Let's simply agree that such a distinction can be drawn, because even if it can be, constructivism doesn't think it licenses the restriction on belief that van Fraassen suggests. The reason for this again illustrates a natural fit between what constructivism says, and how we actually practice science. It fits the circumstances under which we decide that an *observation* has been made, rather than something else – an illusion, a hallucination, or a poorly executed measurement technique.

10.1 Van Fraassen's paradox

The paradox is this:

> Conclusion: The more information a theory contains, the less likely it is to be true. Theories including unobservables always contain more information than the observations justifying them. Therefore, we cannot have a better reason to believe what a theory says about unobservables than we have to be agnostic about them, and believe only what it says about observables. (van Fraassen in Earman 1983, 165–176)

The argument for the first premise begins thus:

1. If a theory T provides information that T^* does not provide, and not conversely, then T is no more likely to be true than T^*. (van Fraassen in Earman 1983, 166; Popper 1959, 399; 1963, 218; and, originally, William James)

The interesting case, which I will presume from now on, is where T is a full theory making claims about unobservable and observable objects, properties and processes, and T^* is that part of it that only makes claims about the observable. If we accept subjective Bayesianism even as an account of strictures on belief that ought not to be violated, this seems completely secure. Anyone's subjective belief in $(P \wedge Q)$ cannot be greater than their belief in P. If it were, we would get a violation of both Bayesianism and common sense. No one can think that two events or states are more probable than one of them.

This depends upon the premise:

2. A feature of T cannot provide more reason to believe that T, as opposed to T^*, unless it makes T more likely to be true than T^*. (Van Fraassen in Earman 1983, 167, slightly altered)

This is uncontroversial, though, for if one believes something one holds it to be true. That is what a belief *is*. The only obscurity is what 'a feature' is, and this gets cleared up when 'features' are picked out by the application of the argument to hypotheses of rival theories.

The next premise is somewhat controversial, and I will take its formulation from another of van Fraassen's works. It is what he says is at the heart of empiricism, and which might be called the *empiricist premise*:

3. Observation is our sole source of information about the world, and its limits are strict. (van Fraassen in Churchland and Hooker 1985, 253)

Thus, the justification for T is T^* and T^* is a proper part of T. Although the strictness of the limits of observation is controversial, the idea that only the outcomes of observations can justify scientific claims is not (*pace* Brown, 1991).

Van Fraassen first stated the empiricist premise in terms of experience, rather than observation. By 2002, he had repudiated this version (2002, lecture 2; 2007, 366–368). Still, he sees empiricism as a stance that takes discovering empirically adequate theories of the world to be the aim of science (2002, 200).

Van Fraassen makes it clear that what I have called the empiricist premise is not a doctrine or dogma (2002, 42; 2004, 172). It is a practical commitment, a stance. In this case it is a commitment to approach the subject of science in a way that does not accord authority to anything except observation. Van Fraassen compares the difference between a stance and a doctrine to the difference between holding utilitarianism as one's ethics and considering it as a philosophical position (2004, 172–174). All the same, whether it is a stance or a doctrine, the upshot is the same:

4. No empiricist can hold it to be rational to believe what a theory says about unobservables, as opposed to restricting belief to observables, and being agnostic about unobservables.

That is the conclusion of van Fraassen's paradox.

10.1.1 Why this is a paradox

This is a paradox in the sense that unobjectionable premises appear to force an objectionable conclusion upon us. If van Fraassen's argument is compelling, then as he notes, we cannot ever be better justified in believing a theory as a whole as opposed to believing only in what it says about observables. We can remain agnostic about what it says about unobservables, and our doxastic attitudes will be at as great a risk of refutation, without incurring unnecessary philosophical problems (van Fraassen 1980, 1983, 1989, 2001).

There is a second reason to regard the argument as paradoxical in the work of Mitchell (1988), Horwich (1991) and Blackburn (2002), which I will briefly rehearse. We just do not seem to be able, when we practice science, to give the agnosticism van Fraassen recommends any real bite. It looks suspiciously as though we really do believe hypotheses about unobservables, but won't say that we do.

The argument goes like this. Suppose a scientist, A, reads van Fraassen, and becomes a constructive empiricist, while another B, continues to believe in unobservables. Suppose A and B are in the same field, working on the same projects.

A is (apparently) agnostic about hypotheses concerning the unobservable, and *B* believes them.

But what *difference* does this difference in attitudes make to the way they both do science? They both draw inferences using hypotheses about the unobservable, use those hypotheses in experimental design, and cite them in explanations, grant applications, etc. If *U* is some hypothesis about unobservables, then certainly they will *say* different things about their own attitudes; *A* will say he's agnostic about whether or not *U*, and *B* will say he believes that *U*. But this difference in utterances looks to be the only difference between them. In other contexts, when people change from belief to agnosticism, it typically changes the way they act; they do more double-checks on things, and do not casually bet their time and energy on projects that would be pointless if the hypothesis is false. But these changes are absent in this instance.

I said that the conclusion to van Fraassen's paradox is objectionable. Why is this? One source of objections derives from Scientific Realism. Under that view, the aim of science is to tell us the truth about the world. Scientific theories cannot avoid making claims about unobservable entities, as the Logical Positivists hoped they could. So science must include the aim of telling us the truth about unobservables. But an argument favoring agnosticism towards hypotheses about unobservables is an argument that we should not take science to have this aim.

At the heart of the paradox is a relationship between evidence and truth. A theory with greater content cannot be more likely to be true than one with a lesser content. Whatever else we say about degrees of belief, we believe something if and only if we hold it true. So van Fraassen's recommendations concerning belief are entirely reasonable. And yet they seem troubling.

10.2 The reply to the paradox

What is it about *observations* that makes them so epistemologically convincing? Humanity did not always privilege them this way. There are many other events that happen to humans that carry conviction about the way the world is, and which we have learned to distrust. I mean such things as having a hunch, or seeing a hallucination, or dreaming, or witnessing an illusion – any sort of event that is unreliable as a source of information about the world. What do we see when we see that such sources of information are unreliable?

Constructive Empiricism cannot evade the burden of saying what it is that distinguishes observations from illusions. For van Fraassen, entities are observable and unobservable, and the human body is a kind of measuring instrument (1980, 16–17). This characterizes the privileged things in a way that avoids direct reference to human experiences. Jennifer Nagel (Nagel 2000, 357) cites the point, made by Psillos, that the human animal doesn't possess the ability to discriminate observable entities at all (Psillos 1997, 371; Nagel 2000, 367). Observable objects persist when unobserved. They are three-dimensional. We possess only the ability to discriminate observable entities when they are actually present under suitable conditions. We can detect only some features or parts of them at a time. Nagel urges that van Fraassen needs to say how to justify hypotheses such as (for example) the enduring presence of the unobserved tree in the quad.

Events of observing are not distinguished from illusions/dreams, etc. by the fact that they are caused, for clearly dreams, illusions, hunches, etc. are caused (van Fraassen 1995, 69–70). Privacy is only limitedly helpful. Hallucinations and dreams are private, and observable objects are public. But illusions are often public too, and mirages. Nor is it helpful to say that reality itself distinguishes observations from dreams, mirages, etc. To say this is not to explain *why* observations pick out reality, and why they alone do so (van Fraassen 1995, 70; Nagel 2000, 358). Obviously, sense-data are hopeless as a starting-point (van Fraassen 1980, 72). Human history doesn't suggest that observations are automatically viewed as more reliable than the other category, or in any way self-certifying. Many humans once took dreams to be informative about future events. Nor is it at all clear that observations are identifiable by their individual content. People mistake dreams for observations, and illusions and mirages are even more misleading.

Van Fraassen appeals to science itself to tell us what observable entities there are (1980, 56–59). This will not address the question that Nagel raises. Unless we can tell what observation is, it is impossible to discuss what it can disclose, and what attitude we ought best to bring to it (van Fraassen 2007, 369–370). Once we possess (as Nagel puts it) a framework of publicly observable objects, we can know how to treat scientific theories. But van Fraassen doesn't say enough about how we get this framework (Nagel 2000, 368).

10.2.1 What is different about observation

I will argue that what distinguishes observation from illusion is success in the practice of fault-tracing.

What distinguishes the category of observation is that the events in it have, when combined with the hypotheses that they confirm, predicted other events we also call 'observations', which conjointly justify a single, mutually supporting group of hypotheses. What distinguishes the category we call 'illusions' (and related misleading events) is that they are unsuccessful when used in this way. They cannot be taken to justify hypotheses that predict other indications that agree in conjointly speaking for a single set of mutually reinforcing justifications.

John Locke put the matter much more simply (Locke's *Essay*, Book IV, chapter xi). Our senses *bear witness to the truth of the reports of each other*. Locke emphasizes that the different senses concur. We can smell as well as see roses, and feel the heat of a fire we see. These are observations made with different sense-organs. Just as trenchant, according to constructivism, is the fact that observations of the same modality *bear witness to each other's report on different occasions and under different circumstances*.

Humans occasionally conclude that they must have dreamed an event that they thought they remembered. How do they come to know this, given that dreams give us pleasure and discomfort and so many of the other earmarks of observation? Clearly, because other experiences belie the content of the dream. I thought I remembered getting the dent in the car door fixed. But that doesn't agree with what's in the garage, or with what my wife says. But it seems a pretty vivid memory. Perhaps I dreamed it. Other examples of using observation to detect whether or not some event is the making of an observation are familiar. Macbeth's dagger and the bent-stick-in-water illusion reveals itself by the mismatch between touch and sight.

How did we arrive at the conclusion that rainbows and other examples of what van Fraassen calls public hallucinations should be deemed illusions (van Fraassen 2008, 101–105)? Well, taking them to be veridical conflicted with other observations, and we preferred the latter observations because they were in turn supported by yet other observations that all concurred in supporting a single body of hypotheses. There are many other examples with this public nature. We formerly took many myths to be authoritative. It took time and effort to gather the observations and reasoning that eventually led us to conclude that Genesis 1–3 was not a literally true account of the origins of humans and the universe.

How do we in fact show that something is an illusion, or a hallucination? We fault-trace to show that the hypothesis we would ordinarily take to be confirmed is false. The Müller-Lyer illusion is revealed when we use a ruler to measure the lines. The distance to a mirage is longer when measured by walking than it appears. That is, we use hypotheses and other observations to show that an event that is normally reliable is not under these circumstances. Because it is not reliable, but is otherwise superficially like an observation, we call it an illusion or hallucination.

What about dreams and imaginings? Here, Anthony Quinton put it better than I could:

> Why, as things are, do we have this ontological wastepaper basket for the imaginary? Because, approximately, there are some experiences that we do not have to bother about afterwards, that we do not, looking back on them, need to take seriously. Dream-events, where they have consequences at all, do not have serious consequences. If I dream of cutting somebody's throat my subsequent dreams will in all probability be entirely unrelated to him and to my act. Even if they are, when I am hailed into court I am as likely to be given a bunch of flowers as a death-sentence. (Quinton 1962, 144)

Quinton imagined a series of coherent dreams, continuous from night to night, in which he lived beside a lake, in a fishing community:

> ... beside the lake there is a place for prudence, forethought and accurate recollection. It is an order of events in which I am a genuine agent. There is every reason there for me to take careful note and make deliberate use of my experience. Reality, I am suggesting, then, is that part of our total experience which it is possible and prudent to take seriously. (1962, 144)

Thus, we in fact use the capacity of observations to attest to each other to differentiate observations from illusions, and from dreams.

The capacity of observations to attest to each other *via* hypotheses they confirm is what makes them worth noticing. The ability to fault-trace and independently confirm are a central part of what makes observations informative about hypotheses which reliably predict other observations. It is prudent to attend to many of these hypotheses because they are the only reliable indicators we possess for triumph and disaster. Triumph and disaster are certainly observable, and hypotheses more distantly related to our interests also have the potential to cohere with each other and observations.

If I'm right, then what is superior about observation, as a source of information, is that *different occasions of it can be brought into common accord*. Different outcomes repeatedly and reliably attest to common hypotheses. We can sometimes alter the scope of the experiences we count as observations, for example by declaring some to be illusions. But there are limitations on our ability to manipulate these diverse indications. Under favorable conditions, different sets of observations are indicative of a common hypothesis in a way we cannot avoid. (That is, we cannot trace a fault away from it, and it has massive, interconnected, independent confirmation.) Each of the independent constructive justifications would continue to support the hypothesis alone, even in the absence of the others.

So there are two kinds of belief-inducing events that happen to humans. For one kind, different sub-collections are indicative of single hypotheses, which can be criticized and modified by using other sub-collections and which sometimes successfully predicts still other sub-collections. We call this kind of belief-inducing event 'observations', and we say that the beliefs it induces are reliable, or confirmed. We distinguish it from the other kind of belief-inducing event because the other kind cannot be brought to collude this way. This other kind we call 'dreams' or 'visions' or 'illusions' or the like. That is how we distinguish what it is to make observations, and why we care about doing so. Only when we have done this can we go on to distinguish observable from unobservable entities.

10.2.2 The problem with van Fraassen's paradox

If this is correct, fault-tracing must be acknowledged *by van Fraassen's Constructive Empiricist* as having a special claim to establish truth, since otherwise we cannot distinguish observations from fantasies, dreams, hunches, etc.

Fault-tracing *works just as well for hypotheses concerning the unobservable as it does for hypotheses concerning the observable*. So we cannot say that we will adopt different attitudes to hypotheses about observables over unobservables – the justifications work in exactly the same way.

For van Fraassen, unified support of a single hypothesis from diverse experience is a pragmatic virtue. A pragmatic virtue does not provide a reason for thinking that the hypotheses of a theory are true (1980, 4). But if we are to distinguish observations from dreams and illusions, then unified support from diverse experience must provide a reason for thinking that hypotheses are true.

That is what's wrong with van Fraassen's paradox. It helps itself to a distinction between observable and the illusory. Nobody can actually distinguish between these two unless they agree that the support of a single hypothesis from diverse observed outcomes is an indicator of truth. But then it says that unified support from diverse experience must be a pragmatic virtue, and cannot be indicative of truth. When we allow diverse support to be indicative of truth, though, the apparent contrast between justifications for hypotheses concerning observables and unobservables vanishes. Although observation remains our sole source of information about the world, and although some entities and not others are observable, the way outcomes of observation provide information about the world indifferently informs us about both observable and unobservable entities.

We can make outcomes and our abilities to detect them the subject of investigation and justification. We get a circle, but a virtuous one, not a vicious one. Some apparent outcomes which we call by words like 'hallucination' cannot be made to square with others. We are in fact, though, able to find a rough set of events that, when accompanied by the reasoning they suggest, reinforce each other as events we have detected in the past.

We can also look at the kind of ways we detect events, for example our sense organs. We are successful in analyzing the way our sense organs function, and so far not at all successful in linking this to many of our judgments. (E.g., we know that there are trees, and how they interact with light, etc., and stimulate our eyes. That is been successful as a study in natural science. But we know almost nothing about what then happens that leads us to say "I see a tree!" and similar sorts of mental processes.) So we can describe ways of identifying the kinds of events we can observe, based on the kinds of events our body is capable of being affected by (e.g., these events reflect or emit visible light).

We should remember that the judgments we arrive at cannot be reduced to these detectable processes. I think we can see that an animal is being aggressive. These events must reflect visible light, or we couldn't see them. But obviously we cannot somehow analyze acts of aggression into some particular class of reflections of visible light.

So it's a non-vicious kind of mutual confirmation. We can take different points in Ohio as starting-points, and wind up with the same hypotheses being confirmed about maps of Ohio. We are not stuck in a cycle of justification here. We can always find new techniques for mapping Ohio, and we know this.

11 Human Values Are Irrelevant to Empirical Justification

What would this change about the way we see science, and the philosophy of science, if one were convinced of constructivism? It would lead to a shift away from pragmatic virtues generally, and social and political involvements in particular, in the philosophy of science. It would allow us to see science and the philosophy of science as much more authoritative in philosophy, and contribute to a more skeptical view of metaphysics as a discipline autonomous from epistemology.

11.1 Why does Quine-Duhem matter to philosophy?

The usual answer to this is that Quine-Duhem undermined the analytic-synthetic distinction in the philosophy of language, and showed that foundationalism was hopeless. Constructivism shows that we may accept the falsehood of foundationalism and suspend judgment about whether the analytic-synthetic distinction is tenable without agreeing to the Quine-Duhem hypothesis.

But Quine-Duhem had a much wider influence on philosophy, and established many positions that constructivism either explicitly repudiates or throws into doubt. As I see it, the most serious of these are:
1. The Quine-Duhem thesis prevents us from seeing that just part of a scientific theory may be justified by the observations.
2. It obscures the idea that observation can be decisive even if it is always theory-laden.
3. It prevents any understanding of independent empirical justification.
4. It denies the efficacy of observation in fault-tracing.
5. It supports the role of pragmatic virtues as indispensable for deciding what hypotheses are best justified empirically.
6. It insulates ontology from concerns about epistemology, and insulates metaphysics from physics.

We have seen the way in which constructivism challenges positions 1–4. In a much less detailed way, this chapter describes how constructivism can provide an alternative viewpoint on 5. The next chapter argues broadly for 6.

11.2 Scientific change does not require pragmatic virtues

The philosophy of natural science has inherited two views that are in some tension with each other from the work of Quine, particularly as it features in Thomas Kuhn's *The Structure of Scientific Revolutions* (1962 [1996]). On the one hand, there is the Quine-Duhem hypothesis. On the other, the view that science proceeds within an overarching consensus, which Kuhn referred to as a paradigm. Kuhn's notion of a paradigm meant that natural science must share a widely agreed upon framework of theories, methods of experiment, problems to be addressed, measurement techniques and instruments, technical vocabulary and concepts, metaphysical presuppositions, and much else besides. These provide the broad commitments "without which no man is a scientist" (1962 [1996], 42). Paradigms are essential to the scientific enterprise: "there is no such thing as research in the absence of any paradigm. To reject one paradigm without simultaneously substituting another is to reject science itself" (1962 [1996], 79).

Under the vast majority of conditions, a single paradigm is dominant in any scientific field. There are times of crisis, when more than one paradigm is competing for supremacy, but Kuhn makes it clear that these are unusual and temporary. A successful paradigm serves to insulate its practitioners from even the perception of alternatives. Paradigms provide a set of problems to be addressed, and methods for addressing them (1962 [1996], 10). Failure to find a suitable fit is a research failure, which reflects upon the scientist, as opposed to the paradigm (1962 [1996], 35).

According to Quine-Duhem, subsets of present scientists could have held onto alternative hypotheses, if they had consulted only the evidence. The result would have been a scientific practice that included a great diversity of opinions. But they did not, and do not, hang onto hypotheses in the light of any evidence whatever. One single viewpoint normally holds sway, even if it can only be roughly described.

The obvious conclusion for Quine-Duhem is that scientists do not cater their beliefs only to the evidence. Other kinds of values forge the consensus on this view, so that those who deviate from the dominant paradigm are not taken to have justified beliefs. This uniformity of opinion has endured in natural science for at least many hundreds of years, perhaps thousands. According to Quine-Duhem, it could not have occurred if scientists had used justification by the outcomes of observation as the sole value according to which they adjusted their beliefs, for clearly that value alone would result in diversity among scientists. They must possess pragmatic values in addition, which they regard as part of justification.

11.2 Scientific change does not require pragmatic virtues

Pragmatic virtues will be familiar by now (van Fraassen 1980, 1985). A pragmatic virtue of a theory is any advantage the theory enjoys besides support from observations. For an empiricist, for whom observational support alone is indicative of truth, a pragmatic virtue will be some reason to pursue a theory that is independent of whether it is true. Explanatory power and simplicity are examples of pragmatic virtues. Like van Fraassen, my objection to pragmatic virtues is that I can see no reason why theories that possesses them should be true (van Fraassen 1980 4, 87, 88; 1985, 285, 295). If a theory is simpler than its rival, or more convenient to test, then that is clearly some reason for working on the theory rather than its rival, but it is not a reason to think that the theory is true as opposed to its rival.

One thought that constructivism opens up, then, is that there is much less need to account for the uniformity of opinion in the natural sciences by introducing pragmatic virtues or conservative thinking about theory change. One might instead explain the uniformity by the greater support that orthodoxy has, or can be reconstructed as having, from the evidence available at the time. Of course, scientists have sometimes been subject to biases – they're human beings. But they have usually done their best, at least over the last century and a half, to be guided in their beliefs by the outcomes they knew about and the reasoning about what it supported.

It has been very common to suggest candidates for these pragmatic features in the philosophy of science. In "Two Dogmas" Quine suggested that both *simplicity* and *conservativism*, that is, the desire to disturb the system as little as possible, guided the choices we made when we modified our beliefs (1953 [1980], 44, 46). Imre Lakatos also endorsed conservativism. He identified a rough core of a research program that scientists tried to save in the face of the data. When saving this core required many frequent, deep, and unproductive changes, the program tended to be abandoned if a rival was available (1978, 34). Many other authors have nominated various values. Hempel, influentially, listed scope of application, novelty of tests, lack of *ad* hoc elements, theoretical support, simplicity, and plausibility given what else is supposed (1966, 28–46). Larry Laudan favors whether it is *rational* to hold onto a hypothesis come what may (Laudan 1996, 34), and then includes in his account of scientific rationality such things as *generality*, *orthodoxy* with the prevailing tradition, and addressing its favored problems (1996, 77–87; 1977). Ian Hacking speaks of *robustness* as guiding the actual choices that must be made (1999, 71–72). Peter Achinstein draws upon the *explanatory connection* between evidence and hypothesis to make the connection (2001, 231–234). Philip Kitcher concedes the logical point of Quine-Duhem, but emphasizes particularly the *unification* of theories in accounting for a wide range of phenomena (the *scope* of the theory) under a single explanatory scheme. He, too, includes a nod to conservativism in the publicity and prior *history of scientific practice* to select among hypotheses (1993, 249–252). Susan Haack goes further than any other author in denying that science is beholden to pragmatic virtues, and claims that social values, at any rate, are irrelevant to its projects (1998, 104). Her extended treatment of scientific investigation on analogy with a crossword puzzle, though, acknowledges that *explanatory integration* is crucial to the process of arriving at belief (1993, 214, 216–217).

Authors do sometimes say that scientists are not permitted to believe anything they like. Laudan and Leplin, for example, attack Bloor for thinking that reasonable scientists could believe what they like independent of the evidence (Laudan 1996, 52), and deny that all theories are reconcilable with any body of evidence (Laudan 1996, 32). But it turns out that the reason for this apparent denial is that there are reasons other than refutation and confirmation by observation that restrict the hypotheses a reasonable scientist may hold. They are explicit that ". . . the theoretical preferences of scientists are influenced by factors other than purely empirical ones" (Laudan 1996, 52). It is clearly their intention to hold that this influence is inevitable, and that we cannot practice science without it, and it is this that I wish to deny.

Such brief references cannot do justice to the sophisticated positions of these authors and others that I have not mentioned, but there is justification enough here for my point. None of these authors allow that a consultation of what the community has actually observed at the time would, by itself, be enough to figure out what should be retained and what amended in a scientific theory.

There is a complexity here, for some virtues that have been called "pragmatic" are correlated with better chances of constructive support from observations. The scope of a theory is an example. A theory with a wide scope, for example, applies to many kinds of phenomena rather than just a few. Hypotheses of theories with wider scope may be more likely to get a wider range of independent support from different sources when contrasted with hypotheses from theories of more limited scope. So, over time, scientists who are interested only in which hypotheses are best justified are likely to display a pattern of proposing theories with wider rather than narrower scope. They will tend to believe hypotheses of wider-scope theories as opposed to narrower-scope ones, not because of that scope, but because it carries with it better-justified hypotheses. There is an analogy to biology here. Life displays many examples of apparent self-organization and design. But that is not because self-organization and design are biological forces directing the history of life. They are correlated with what does do so.

11.3 Justified empirical beliefs do not depend upon social values

Writing of Paul Feyerabend, Sandra Harding noted:

> The consequences of the Duhem-Quine thesis have been extended [. . .] to open the possibility of a reconsideration of the link between epistemology, on the one hand, and ethics and political theory, on the other. (1976, XXI)

If constructivism is so then these consequences do not follow for natural science, and so far as justification is concerned, social and moral values are irrelevant to it.

Justification from past outcomes of observations does provide a reason to think that a hypothesis is true, and for an empiricist, nothing else does. So while you might *choose* to believe for pragmatic reasons, including moral, political or social values you hold, empiricism says that your belief is not justified in such a case, because there is no connection to truth in your reasons for belief. Constructivism argues that we can get on with the practice of science perfectly well using only the past outcomes of observation we have detected and recorded. Only by doing so are our beliefs justified, because only outcomes of observation are indicative of empirical truth. That is the sense in which constructivism says that moral, political, or social values are irrelevant to natural science.

Perhaps some radicals will reply that their political commitments override the value of believing those hypotheses that are best justified. Or perhaps, less radically, one could point to the fact that support by observations, no matter how strong, can always be defeated by future evidence. So one could cite moral and social values to fix belief, and hold that the present evidence is misleading. One might also hold that it is less bad to refuse to believe in the teeth of the present evidence than it is (for example) to adopt a belief that is somehow pernicious. (This argument supposes, evidently, that we can avoid believing a hypothesis that is better supported by the outcomes of observation if it is sufficiently politically repulsive.)

If constructivism is accepted, then we should hold that, although all kinds of social and political factors might in fact have influenced the course of the history of science, this need be of no interest to a working scientist in guiding belief and disbelief. Certainly social, political, and other broadly moral factors are relevant to issues other than whether a hypothesis is justified. They are relevant to, for example, which problems we choose to address, and the degree to which a result appears significant to us. They are of relevance to the availability of funding, and to the intensity of criticism. Helen Longino argued that a diversity of social backgrounds for scientists would be a resource upon which scientific practice could draw in promoting a diversity of empirical theories and critical viewpoints (1990, 62–82). Constructivism need not dispute that. Social and political values are relevant to science in a great many ways. Constructivism, though, argues that they are irrelevant in one specific way: many have argued that observations by themselves are powerless to get us to believe anything, so that the intrusion of social, political or moral factors is inevitable in taking hypotheses to be justified (for example, Harding 1976, XXI; Kuhn 1977, 330–339; Longino 1990, 59; 2001, 50). Constructivism disputes both the premise and the conclusion here; outcomes of observation are quite sufficient to show why we believe the hypotheses we do, so that we can fully account for why we ought to believe the hypotheses we do in natural science without the intrusion of social and political values.

That point is consistent with the need for social and political values in other contexts. Along with everyone else I hold that social values will be relevant in deciding, for example, whether and how actually to perform experiments. There will always be a need for human subjects committees. Nor am I defending the view that theories or hypotheses are proposed without regard to human interests. All I argue is that which outcomes of observation actually occur are often too stubborn to be budged no matter what our human values, and that only beliefs that are guided by which outcomes occurred are justified. What is empirically true or false is indifferent to our human values.

This opens scientists up to serious moral challenges. It is possible to encounter evidence that an assumption behind some moral, political, or religious commitment one has is false. While these challenges are most common in social sciences, which I have not discussed, they arise from natural sciences too. In every age there are facts about the world, about our origins or our biology, that are entangled with our moral viewpoints. In every age it is regarded as immoral to believe things about the world that are incompatible with these viewpoints, or even to be agnostic. If the evidence challenges such a viewpoint, there is a deep temptation to resist it and its implications. There is a good case for saying that to a degree, that is what we ought to do, because we are moral beings as well as scientific investigators. Even if we ought to resist evidence to some degree, though, it is a serious and difficult question what that degree is. At some point, since we cannot save the hypothesis come what may, the evidence can present a profound moral dilemma. Are we resisting too far? Not far enough? How are we to think and act if we surrender the hypothesis? Ought the evidence to be suppressed because of its probable deeply pernicious effects?

The prejudices of our time lead us to look down on those who opposed the motion of the earth, or the biblical view of the history of life on earth. It cannot have been easy, though, to have embraced the contrary as a deep part of one's life, and then been exposed to the evidence against it. It isn't impossible that we still might face the kind of problem that Robert Cardinal Bellarmine and the Reverend Charles Kingsley faced, although the context will be quite different.

11.4 Does constructivism really avoid pragmatic virtues?

One might argue that constructivism depends on a pragmatic virtue as much as any view of empirical investigation does. It reconciles conflicts between the many constructive trees we are capable of making up by looking for those hypotheses that possess most justification from most sets of outcomes of observation. So constructivism is not *just* looking at whether or not a hypothesis fits the outcomes. It is also looking for a unified perspective: a single set of hypotheses that the outcomes of observation all support together.

Why, according to constructivism, do we not believe the phlogiston theory and vitalism? Well, because there are few, constructive trees confirming the hypotheses of these theories, and those there are suffer serious problems with some of the outcomes of observation. By contrast, there are many constructive trees supporting rival theories, and these do not run into trouble with other outcomes we have observed. But you might argue that phlogiston and vitalism only suffered these ignominies because they were being evaluated by someone who endorsed the pragmatic virtue of unification. If we drop that pragmatic virtue, there's no need for justified entities to have a very wide range of application. These entities and their theories need not fit well together; we might have a kind of patchwork quilt of theories, each with a very limited application.

The idea is not an idle one that nobody has ever considered. Van Fraassen entertains such a picture:

> . . . surely it is possible to have a lot of theories, each with its individual sorts of models, more or less overlapping in their domains of application – all empirically adequate, but impossible to combine into a single picture? (1980, 86)

He goes on to point out that there are domains where the contrasting theories overlap, so that the idea is "really not feasible" (1980, 86). But as he also points out, there are cases in real science where just such a situation obtains. Physiologists do not make relativistic corrections in their mechanical calculations (1980, 87).

The first thing to say, in reply, is that in spite of the cases van Fraassen cites, the natural science we actually possess does strive for theories with a wide range of application in the way that constructivism says it does. We do not favor just anything that gets any initial evidence, and then preserve it by restricting its domain. As a general principle, that would result in a science we simply do not observe. When we do restrict domains and use theories that we know are false, we give reasons. The reason physiologists do not make relativistic calculations is that we have good evidence that there is no point in doing so because the corrections are well below the threshold of experimental error.

But that reply can clearly be taken to make the problem even more serious. It isn't conflict with the evidence that prevents a chaos of differing opinions in natural science, the objector will reply. It's just that contemporary natural science does endorse the pragmatic virtue of unification. There is, intuitively, a very strong pull to the idea that independent justification adds additional justification. The objection claims that this intuitive appeal is simply a reflection of our fondness for this pragmatic virtue. There is no consensus at present on how exactly different independent justifications *quantitatively* add to justification. Thus objectors will feel confident saying that it is not conflict with the evidence that prevents us from having a patchwork quilt of theories in the way van Fraassen entertained.

The best answer to this objection, in my view, is that unification in the constructive sense is not a pragmatic virtue. A pragmatic virtue is one that leads us to prefer a theory whether it is true or not. But to the degree that we can discover empirical truth at all, we do so only by seeking sets of hypotheses with a unified justification from the outcomes of observation. There is no way to recognize one of a set of hypotheses to be empirically true besides their collaborating in independent justifications of each from the evidence, with no similarly successful refutations or any of them. That is:

Some kind of unification of independent justifications without refutations is a necessary condition of empirical inquiry.

Call this *the broad unification condition.*

Constructivism presents an account of this kind of unified justification. It is to be able to come up with some constructive trees justifying the hypothesis, and none, or only doubtful ones refuting it. More such justifications from many independent sources makes the justification stronger. The justification is also strengthened when the auxiliaries used in these trees have very strong justifications that are independent of the target hypothesis, and when the independent backing-hypotheses for the observations are strong. Perhaps other accounts besides constructivism are also feasible. But so far, constructivism is the only one we have got.

There are a number of overlapping reasons to believe that the broad unification condition holds:

First, it's simply very difficult to think of any examples of hypotheses we do take to be true as a result of empirical inquiry that do not have a lot of unified evidence in their favor. We might not know exactly why and how a commercial pregnancy test works, but we trust it only because we know it must often have got the right result very frequently, and couldn't have done so without an enormous amount of thoroughly checked research. Some results of natural science, such as the detection of the Higgs boson, are sometimes reported as if they were a single experiment, without independent support for the hypotheses involved. But even a casual investigation reveals that this "single" experiment involves a huge complex of evidence to support the auxiliaries. There are experiments to establish that the initial conditions hold, that the apparatus is correctly designed and set up, and that the detection instruments work. Then there are other experiments that support the interpretation of the observations. Only after this is complete are the highly selective and cultivated outcomes of observation taken to be evidence for the hypothesis. The auxiliaries have to be rock-solid before "the" experiment can take place, or we wouldn't accept its conclusion.

Second, there is the problem of identifying observations. It's universally allowed that whenever we engage in empirical inquiry we make observations. If even being able to do *that* entails unification, then engaging in empirical inquiry is sufficient for accepting unification as truth-indicative. But if we are to be able to distinguish between an observation and some misleading mental event, then we have to check different results against each other to find out which are misleading. That is, we have to unify the suggestions prompted by different mental events to arrive at mutually confirming methods of observation.

We covered this in chapter 10, of course. Illusions, dreams and hunches must be differentiated from the effects of physical entities on our senses. The effects of such entities on our senses is highly informed by initial conditions and training. Only some kinds of equipment, under some circumstances, are acceptable. Even without equipment, we must be trained to observe. And of course, we do not conclude we have observed something in the absence of backing hypotheses concerning the observed object, property, process, event or state of affairs. So even deciding what counts as an outcome of observation, under what circumstances, is a matter of integrating a great deal of evidence. (Perhaps even non-human animals do something similar, for they appear to learn that some stimuli are misleading.)

Third, there is the point, repeated throughout this book, that if we are to discover any empirical truths, then we must fault-trace. Fault-tracing uses the results of different outcomes to arrive at a body of hypotheses that are all confirmed together. So again, if we are to arrive an any justified conclusions, we must integrate different outcomes of observation.

Fourth, and relatedly, in empirical investigations we independently confirm a single hypothesis by multiple methods. These cannot be different if they do not depend upon different outcomes, and they cannot agree unless there is a single hypothesis for which they all speak.

Fifth, look at any overwhelmingly confirmed hypothesis. In these cases, maintaining that the hypothesis is false marks you as quitting the science game.

Well, why? Because each attempt to get the hypothesis to be false conflicts with outcomes of observation that are impossible to evade. There must be a number of different ways to try to maintain the falsehood of the hypothesis, and they each run into different problems. In historically contested cases, we settled on the hypothesis we did only because each of these objections was met with a different kind of answer. It is the success of different kinds of evidence, and the separate hypotheses for which they each speak, that forces us to realize that we share a common ancestor with chimpanzees.

By contrast, where we have more doubtful or contentious hypotheses, the debates circle around possibilities that cannot be blocked by outcomes of observation. There was a mass extinction of megafauna in the Americas around the end of the last ice age. Did humans cause it? It is difficult to rule out other possibilities, such as the change in climate, because we do not have outcomes of observation that differentiate between the hypotheses. So the issue is more open.

Sixth, when pragmatic virtues are severed from independent justification from the evidence, they are unconvincing as recommendations for empirical belief. Compared to independent justification from outcomes of observation, there seems little reason why reality should possess any of the pragmatic virtues. The world doesn't have to be simple. It doesn't speak for the truth of some conjecture that it applies in lots of areas if it doesn't run any risk with the evidence in any of those areas. There is some intuitive appeal to the idea that a hypothesis that explains a lot is more likely to be true. But Karl Popper long ago made the point that, although good theories explain a lot, bad theories often explain even more (1963, 33–42). Simplicity, scope and explanatory power look good when they are harnessed to hypotheses that, in company with others, are confirmed by the outcomes of observation. They do not recommend belief just by themselves.

Seventh, support by independent outcomes of observation is needed to substantiate arguments from pragmatic virtues, and in the cases of a clash between the two, observation always overrides pragmatic arguments.

Pragmatic virtues have some power to get us to believe things by themselves. Although empiricists frown on it, we do sometimes change our point of view without the addition of new evidence by citing pragmatic virtues. One day, for example, it might suddenly occur to you that your dog's repeated flea infestations coincide with the visits of a friend who brings another dog to the house. It is simpler to think that the fleas on the visitor's dog repeatedly re-infest your dog than to believe that independent causes of flea-infestation coincide accidentally with the visits of the other dog. So you might change your mind about why your dog gets fleas without any new observations, by adopting this simpler explanation.

The point is that justification by interdependent, independent, constructive justification will override changes in viewpoint brought on by pragmatic justifications. Where there's a conflict between the outcomes of observation providing independent justification and some pragmatic virtue, the former take precedence. Someone who holds onto a simpler view in the face of contrary evidence is regarded as (at least) taking an epistemic risk. Someone who holds onto a complex view that the evidence better independently supports in the light of a simpler view that is not as well supported is just being sensible. If you had good evidence that the visiting dog had never had fleas, it would be daft to continue to believe that the re-infestations were due to its having visited.

Other virtues besides simplicity get overridden by evidence. We do not sacrifice better independent support when we are presented with a theory with greater scope but less independent support. The independent support dominates the greater scope. If this were not so, we would believe powerful theories that have been empirically refuted, as we clearly do not. All kinds of mumbo jumbo has a scope that is wider than the supported hypotheses that rival it – astrology, quack medicine, conspiracy theories, the idea that God is testing our faith by making the rocks look old, and the idea that everything happens by accident are all examples.

In summary, then, I can see no way out of the idea that we if we have a justification for a hypothesis of natural science, or even a mundane empirical hypothesis, we have integrated many sources of support from different outcomes of observation. Constructivism is one account of the way that we do this. There might be another account, or even – perish the thought – a better one. But so far, constructivism is the only one available.

Van Fraassen once noted that it is often not at all obvious whether an entity in a scientific theory is mathematical or physical (1980, 11). The line of thinking defended here suggests a way to distinguish them. In the case of hypotheses about the empirical world, justification requires integrated support from different outcomes of observation. The hypothesis that some entity is physical rather than mathematical, then, is safe if it is actually supported in this way. By contrast, anyone who has a good reason to believe that the existence of an entity, or kind of entity, cannot be justified in this way, ought to be criticized if he or she says that it is physical, and explains or justifies things that are physical. Still, it might be a mathematical entity. That doesn't take us very far with van Fraassen's puzzle. All the same, I believe it is useful in clarifying, for example, what quantum mechanics says about the physical world, and where it is silent about it (Mitchell 2019, 805–806).

12 From Constructivism to Metaphysics: Potential Applications

The Quine-Duhem hypothesis lies deep within contemporary philosophy and profoundly affects the way philosophers see the world. It is, as Clark Glymour once put it, the lynch-pin on which philosophy turned in the mid twentieth century (1980, 5). To give some sense of the potential consequences of abandoning it, this final chapter presents three ways in which constructivism might be applied to metaphysics, to give some sense of the range of possibilities of thought which have been apparently closed by the Quine-Duhem hypothesis. I shall be painting with very broad brush strokes; the point is to show that the argument of this book opens up potential avenues, not to follow those avenues for any distance. That will be a job for future work.

12.1 Empiricism

Philosophy has applied the Quine-Duhem hypothesis in many ways. Some have used it to introduce human values into our view of empirical justification. Others have used it to argue that non-linguistic animals can have no beliefs. Or it's been alleged that as a result of Quine-Duhem, expressions have no sense, but only reference. It's also been used to justify holistic views of mind.

But its most pernicious effect, in my view, is in metaphysics. Quine stated this effect only in broad outline at the end of "Two Dogmas of Empiricism" and took it for granted in the background of his "On what there is" (1953 [1980], 1–20). Put rather vaguely, this is the idea that it is simply impossible for metaphysics to emulate natural science in the sense of seeking value-independent conclusions about what there is and what is true. It is impossible to seek hypotheses that are justified no matter what we prefer. When we adopt a view about reality, or what there is, ". . . it turns upon our vaguely pragmatic inclination to adjust one strand of the fabric of science rather than another in accommodating [the outcomes of observation]. Conservatism figures in such choices, and so does the quest for simplicity" (1953 [1980], 46).

Prior to Quine-Duhem, natural science appeared to be the ideal model of actually discovering what is true about entities in space and time, including ourselves. Quine's argument was directed against the logical positivists, but this ideal stretches back long before that. By beginning with the idea that experience is our sole source of information about the Universe, philosophy emulated natural science and could reasonably hope to establish truths that are just as independent of our preferences and illusions. For when Quine wrote "Two Dogmas", philosophy thought of natural science as a means of discovering truths independent of human preferences or even of our current beliefs.

But Quine-Duhem allegedly shows that this is a naive view, a kind of science-worship. For natural science is as guilty of involvement in human interests and values as philosophy has ever been. There is no reason for the Universe to be simple, rather than complex, no reason for truths about it to be widely applicable rather than narrowly, or minimally different from what we believe as opposed to maximally. These are values *we* find attractive, or convenient. Yet, without them, outcomes of observation reveal nothing. The great tragedy of science, as Huxley put it, is that ugly facts will slay beautiful hypotheses. Under the Quinian picture, human values, desires, preferences, interests– call it what you want – are required for empirical justification. According to Quine, no fact is ugly enough to refute a hypothesis in spite of these values. In this sense, at least, there is no (significant, epistemic) difference in kind between positrons and the gods of Homer (Quine 1953 [1980], 44).

Except that there is a difference. The gods of Homer are a fiction inspired by the beliefs of their day and accompanying human interests. Positrons were a prediction that was compelled by the Dirac equation and subsequently confirmed by photographs of cloud chambers, in spite of our prior beliefs or anybody's interests. The scientific worldview is a system that has to cope with outcomes that it cannot control, and which can over time decisively refute its fondest illusions. You can sneer all you like at Popper and the logical positivists, but what impressed them about natural science remains strikingly impressive: scientific hypotheses take a *risk*. They can be found to be wrong no matter how strongly anyone wants them to be right.

The principle importance of the Quine-Duhem hypothesis, then, is that it destroyed the possibility of a certain approach to metaphysics.

The approach to metaphysics had been to look at natural science to see what it said was so, and to look at the preconditions for natural science to be a genuine method for justifying these claims. Michael Friedman has been prominent in his study of such a view:

> Schlick aimed to do for Einstein's physics what Kant had done for Newton's, namely, to explain and exhibit the special features of this physics that make it a model or paradigm of coherent rational knowledge of nature. One central implication of this new physics, however, is that Kant's conception of natural knowledge, as framed by universal forms or categories of the human mind, taken to be rigidly fixed for all time, cannot, after all, be correct. (2001, 14)

The Kantian categories of the human mind having proven to be flawed, Carnap replaced them with frameworks of knowledge in "Semantics, Empiricism, and Ontology" (Carnap 1956). These were not fixed for all time, but were nonetheless immune from investigation by the framework. They were required in order to get any kind of investigation within it to work.

"Two Dogmas of Empiricism" presented a picture of empirical investigation that made this approach impossible. *Any* hypothesis could be affected by our reactions to empirical investigation, and any hypothesis might be found wanting as a result. So the "framework" was illusory. Nothing was fixed as a presumption of empirical investigation. At the same time, any hypothesis we like can be retained no matter what the outcomes of empirical investigation are. So no hypothesis needs be given up, no matter how the observations come out. So one central moral of "Two Dogmas" was that metaphysics could not begin with natural science and the value-independent hypotheses it justifies. That could not be a method even for natural science, because it cannot justify hypotheses in a value-independent way. For Quine, philosophy was continuous with natural science. Philosophy could not offer some *a priori* grounding for *a posteriori* justification (1969, 126). But this continuity was not natural science providing a human-interest-independent source of knowledge for philosophy. It was rather philosophy revealing that natural science was a human-interest-*dependent* source of knowledge in the first place.

And so an approach to metaphysics developed that saw it as the attempt to strike a balance between different sources of information that maximized various pragmatic virtues, such intuitive appeal, simplicity, and explanatory power. In the words of David Lewis:

> One comes to philosophy already endowed with a stock of opinions. It is not the business of philosophy either to undermine or to justify these preexisting opinions, to any great extent, but only to try to discover ways of expanding them into an orderly system. [. . .] It succeeds to the extent that (1) it is systematic, and (2) it respects those of our pre-philosophical opinions to which we are firmly attached. [. . .] some of us sometimes change our minds on some points of common opinion, if they conflict irremediably with a doctrine that commands our belief by its systematic beauty and its agreement with more important common opinions. (1973, 88)

Lewis is speaking about possible worlds, of course. As Callender (2011, 35) put it: "In our cartoon-like history, we might say that Quine cleared the room for metaphysics, while Kripke furnished it."

The philosophy of science has a long history of suspicion about how much metaphysical knowledge we possess, so it is not surprising that something of a reaction eventually set in from that quarter. The work of Ladyman and Ross is a vivid example. In 2007 they presented a very strong rhetorical case for a return to founding metaphysical investigation upon our best current science – a position they dub 'naturalism'. The advantage of this position over the metaphysics of that day ('analytic metaphysics'), they argued, is that it is better connected to the scientific enterprise. "[E]ven if naturalism depends on metaphysical assumptions", they wrote "the naturalist can argue that the metaphysical assumptions in question are vindicated by the success of science" (2007, 7). The position echoes a more modulated case made by van Fraassen a few years earlier (2002, chapters 1 and 2), by Penelope Maddy (2007), and by Tim Maudlin (2007, Epilogue).

My central objection to this enterprise is that *without a reply to the Quine-Duhem hypothesis, a naturalized approach secures no better connection to truth than does analytic metaphysics.*

Ladyman and Ross's case for the superiority of a variety of naturalism which is based firmly in empirical results depends a great deal on the flaws they detect in analytic metaphysics. The intuitions upon which it depends are epistemically dubious. They vary culturally, and are subject to experimental manipulation (2007, 10–11). They are based on a very narrow range of physical magnitudes (2007, 11). Things are taken as certain which are not actually certain at all (2007, 13–14). On the few occasions when analytic metaphysics draws upon empirical results at all, it avoids anything recent and uses (at best!) empirically refuted theories such as Newton's (2007, 17, 24–27).[19]

Analytic metaphysics has to appeal to virtues such as simplicity, systematicity, scope, and intuitive appeal in order to decide metaphysical hypotheses. Critics such as Ladyman and Ross make some very telling criticisms of the approach. But the criticisms are telling because there is no reason for entities that are human-independent to be simple, or subject to a wide-ranging analysis over a narrow-ranging one, or intuitively appealing. The trouble is that exactly the same objection applies to the pragmatic virtues that Quine-Duhem alleges are essential to naturalism. It is true that natural science ignores virtues like intuitive appeal, so the "virtues" are not identical. But that doesn't matter when both share this difficulty. There are too many contradictory ways of balancing these virtues, with no way to reconcile which is correct. Ladyman and Ross want to throw out the epistemological bathwater of analytic metaphysics. But without a reply to Quine-Duhem, they have no baby to retain instead. They are simply replacing it with a different kind of bathwater.

---

19 Many similar criticisms have been lodged in naturalized epistemology (for example Hilary Kornblith (2014)), and by the experimental philosophy movement (an overview is Knobe & Nichols 2017).

By contrast, once we have an argument against Quine-Duhem, there is a much better case for beginning metaphysics with natural science. Overwhelmingly confirmed hypotheses are the effects of the physical Universe upon our physical senses that we cannot devise a way to evade. They are, then, our best guide to the way things are. The outcomes of observation sometimes constitute an unrivaled mutually reinforcing class of independent justifications in decisive support of some hypotheses. Other approaches to metaphysics do not have a similarly strong argument linking them with knowledge.

The argument is circular, because it is only reasoning from our experience of causes and effects that reassures us that our senses are indicative of the states of other things. So we are using the results of empirical justification to reassure ourselves about its reliability. But this is not guaranteed to succeed, for we need not have discovered that our senses were capable of being viewed as reliable physical detectors. Circularity is vicious only when it cannot fail. And even the structure of our failures is illuminating. Sometimes, outcomes of observation in one area indicate that the hypotheses suggested by outcomes in another area are inaccurate. When the areas disagree, we can often use other observations to suggest a reason why they disagree, and confirm from new observations that it's correct. Quine was correct in citing the empirical vulnerability of Carnap's frameworks. But he was wrong in thinking that there could only be pragmatic tests for where these errors lay.

Nothing guarantees *a priori* that we would find this happy result, and in wishful thinking, conspiracy theories, dreams, astrology and all kinds of quackery, we do not find it. There's a case for verisimilitude in the empirical investigations just by themselves. There is, then, a good reason to look at the necessary conditions for our knowing that overwhelmingly confirmed hypotheses are justified by the outcomes of observation.

Take an example of a metaphysical hypothesis that is a consequence of empirical activity, or of its results, but which we cannot see how to justify. The remarks of the naturalists suggest the following argument. If observations justify hypotheses overwhelmingly, then certain metaphysical hypotheses must be true. Observations do overwhelmingly justify hypotheses, and we know this. So we know that (some) metaphysical hypotheses are true. So this constructive method in metaphysics begins with the fact that we overwhelmingly, although defeasibly, succeed in gaining knowledge *a posteriori*. Since this is so, its consequences are so.

What would metaphysics be if we adopted this approach? I have three broad suggestions. First, if we have the kind of justifications that constructivism says we have, then the knowledge we acquire has some very general consequences which metaphysics ought to examine. I am thinking here of examples like Craig

Callender's examination of the way in which we identify the time dimension and what distinguishes it from spatial dimensions (Callender 2017). This is a thriving area in philosophy, and there are many examples of it in operation. Almost the whole literature of the philosophy of quantum mechanics, for example, consists in attempts to come to grips with the consequences of empirical discoveries in that subject.

These are metaphysical consequences of specific areas of scientific investigation. There are, secondly, more general consequences of empirical methods of justification genuinely delivering knowledge. This kind of metaphysics would include some of the traditional topics in the area – the problem of induction, for example. It would also concern the disputes about the correct method for empirical investigation, for example the thriving disputes in the philosophy of statistical and probabilistic reasoning. This approach to metaphysics begins with the premise that we do get empirical knowledge in the sciences, and thus strongly suggests that the disputes here concern the best analysis of the way we do in fact reason.

Third, there are consequences for us, the human animal, engaged in the activity of gathering empirical knowledge. Empirical investigations are behaviors of ours, actions, and our acting this way confirms various hypotheses about us. Some have argued, for example, that if we reason and direct our activity using unobservables in the same way as we do for observables, it follows that we cannot believe in the one and be agnostic about the other (e.g. Mitchell 1989). There was a tradition, of course, prior to Quine, which alleged that engaging in communication entailed that various hypotheses were true of us, while others lacked any justification.

Under this approach, we cannot draw a precise distinction between metaphysical and empirical hypotheses. It is not as though we know that certain hypotheses, the metaphysical ones, cannot be confirmed. All we are faced with are hypotheses we do or do not know how to justify at present. In 1781, it appeared that the Euclidean nature of space was in the same boat as the persistence of the Universe when unobserved. Both were consequences of our ability to justify empirically, and both were apparently incapable of empirical justification. They were apparently both instances of the Kantian approach to metaphysical theses. But now we know how to test the geometry of space, and we are far more cautious about what the future could reveal. We have learned to distrust the intuition that some hypothesis could *never* be justified, and so *must* just be a consequence of our empirical activity, and *couldn't* be empirically confirmed. The case for saying that a hypothesis cannot be justified, when there is one, depends upon what we have overwhelmingly confirmed. In the same way, the history of refutations should make us very cautious about saying that any hypothesis cannot, ever, be refuted by the evidence.

While I do not want to follow this constructive way of answering metaphysical questions very far, I do want to address an objection. Consider the hypothesis that the Universe continues to exist when unobserved. If one begins with empirical justifications, then the worry is that this metaphysical hypothesis might as well be false. But if it is false, then we do not know, for example, that the arrival of brown snakes in Guam really did result in a crash in the bird population. For it could be that no brown snakes, or birds, exist when we do not observe them, and the new birds that pop into existence when we look are just fewer in number, and the snakes greater, without either affecting the other. If a metaphysical hypothesis is false, then our empirical hypotheses are false too. We know this. So really, we *cannot* begin with the empirical justifications in the way constructivism wants.

My reply is that this argument privileges mathematics over science as a model of justification. The argument says that since metaphysical hypotheses are necessary conditions for empirical justification, empiricism must establish the metaphysics *first*, and only then can it claim that empirical justifications are genuine. That form of argument works well in classical mathematical reasoning. But it's completely the *reverse* of what happens in scientific reasoning. For there, not only metaphysical hypotheses, but utterly mundane empirical hypotheses can be necessary conditions for an empirical justification that are untouched by its success. We do not have to establish these mundane consequences prior to accepting a piece of empirical justification. It's the fact that the observations came out one way rather than another that drives empirical justification. That can still indicate truth, even when it could be defeated if some of its consequences were false.

Suppose I discover that some peanuts have vanished from the kitchen cabinet. Looking around, I discover mouse-droppings, and confirm that a mouse stole them, as opposed to a human having moved them. A consequence of this reasoning is that the inside of the cabinet persists when unobserved. So if that is false, a mouse didn't really take them. But it is not only metaphysical hypotheses that are in this situation. There are lots of far more mundane hypotheses that have the same effect. It's a consequence of this reasoning that I correctly remembered the earlier location of the peanuts, and that there is a way into the cabinet that a mouse can use, for example. (Given the background, it's also a consequence that whatever took the peanuts had a grandmother, and a functioning liver . . .) For some of these, I might have evidence, but probably not for all. But I do not need to *check* them. I have evidence that I can reliably detect mouse-droppings, that mice leave them about, and that they frequently raid human supplies. That, plus the mouse-droppings, is all I need. I observed mouse-droppings as opposed to a note from my wife saying "Delicious peanuts!". That is enough. I do not have to check every consequence of a piece of empirical reasoning in order for it to be an empirical justification.

Still, one could argue that constructivism still gets the analysis of metaphysical hypotheses wrong. The persistence of an unperceived Universe isn't just necessary for justifying empirical hypotheses in the sense of being a necessary condition for them. What makes metaphysical hypotheses metaphysical is their connection to necessity in the sense of *having to be true*, not being consequences of what someone believes. Constructivism has a potential approach here too.

12.2 Possibility without possible worlds

Fault-tracing cannot work without our having the ability to know what would happen if some of the things which we know to be true were false. For, in fault-tracing, we sometimes find out that some of the things we thought we knew to be true *are* false. We couldn't have found that out without being able to reason about what would happen if we were wrong and being able to devise experiments to put such reasoning to the test. Yet clearly, in the present, no mark distinguishes those hypotheses which we know from those which we only think we know. We cannot tell them apart. We must possess a general ability to reason about what would happen if part of what we think of as our knowledge is false, although we might be unable to reason about what would happen if *all* of our knowledge were false *together*.

Fault-tracing is central to constructivism. Whatever constructivism says about necessity and possibility has to be guided by the use of these modalities in this reasoning about counterfactuals. In fault-tracing, there is a data-set we are trying to cope with, and we are trying out the falsehood of different hypotheses to decide which experiments to try, and what would follow from other hypotheses if various candidate hypotheses were false. This search for the error might eventually lead to any hypothesis.

If counterfactual reasoning is to be useful in this way, then the consequents must be hypotheses that are, (or were, or in the future might be), contrasted or matched to physical states. (A way of putting this that is more familiar is "must be contrasted/matched to physical states of the actual world".) That is, in the hypothesis:

If X were so, then Y would be,

we have to understand the consequent this way:

... then Y would *have been* be so, or would *not have been* so.

Y must be something that, detectably, either *actually isn't* but would be, or *actually is* but wouldn't be, if X were so (or not so). The reason is that we have to have the ability to *use* the conditionals to decide what experiments to do or where to observe. We need not know right now which of the two Y is – being so or not – but it's being one or the other is what gives us information about whether or not X is at fault for some difficulty with the data. We get no evidence about whether X is at fault by knowing that Y is "true in some other possible world" – whatever that may mean.

In fault-tracing, I know that X, because I know of lots of data and reasoning from it that justifies it. Still, I face a problem with the data, and I wonder whether X might be false after all. So I try to keep as much of my knowledge as I can, and try to re-conceive things to avoid the justification for X in a way that will solve my problem. That almost inevitably means that I have to try to drop other items of knowledge too. I will probably have to drop auxiliaries in the justification of X, perhaps also some outcomes that justify it, and known consequences for which X is required. So I have to pay attention to only some constructive trees, and set aside those that confirm X, or require that X be so in some way. I will presumably gain new constructive trees that use the falsehood of X.

Sometimes, startlingly, it turns out that I *can* do this. I *thought* I knew that X, but upon rearranging the justifications it turns out I can keep nearly all of the justifications from the outcomes of experience, and account for the justified consequences of X in some other way. Then subsequently, I sometimes get a successful double check on the hypothesis that X is false, and restore consonance. Recall the examples from chapters 7 and 8. It turns out, surprisingly, that Compton didn't in fact know the value of Avogadro's number to the accuracy he thought, because the charge on the electron wasn't known that accurately either. Not all the sunglasses labeled 'polarized' are in fact polarized. The students are not, after all, randomly selecting the beads from the bag. Similar instances of backtracking will be familiar from detective fiction, and from crossword puzzles.[20] So long as the consequent of the subjunctive conditional is something we know of a way to find out, we can use subjunctive conditionals to discover that something we thought we knew is false after all. On other occasions, of course, we do not find this at all. As we'd suspected all along, taking X to be false clashes with stubborn outcomes of observation, reinforced on many occasions and from many different directions.

---

20 Susan Haack drew analogies between science and crossword puzzles, but not quite in this way (1993).

The reason this connects to modality is that different hypotheses resist being taken to be false to different degrees, given our current evidence and reasoning (and similarly for being taken to be true). We can know both hypothesis P and hypothesis Q equally securely, but find it impossible even to get started with the view that P might be false, while conceiving of Q's falsehood, even if ultimately unsubstantiated, has clear and identifiable effects. To use a natural expression, Q's falsehood is *easily conceivable* but P's is not, even though we end up with an overwhelming justification for both. To use another natural expression, P is *necessary* and Q's falsehood is *possible*, although Q is true.

Conceiving of some hypotheses as false changes nearly everything. Not much would still be true if there were no electromagnetism. Florine has atomic number nine. It's very difficult indeed to conceive of that as false. Too much else is bound up with the hypothesis. Thomas Kuhn noted the acute sense of intellectual vertigo that accompanies scientific revolutions, when hypotheses formerly thought to be necessary truths are under active challenge. "All my attempts to adapt the theoretical foundation of physics to this knowledge [Planck's formula for radiation] failed completely", Einstein wrote, "It was as if the ground had been pulled out from under one, with no firm foundation to be seen anywhere, upon which one could have built" (Einstein 1945, 45; Kuhn 1962 [1996], 83–84).

By contrast, a great deal of what we know about the world is still justified if we drop the hypothesis that a particular play in some sports event succeeds rather than fails. It "could have gone either way". Atomic numbers, conservation laws, even taxonomies of species such as the hypothesis that whales are mammals, seem to be necessary truths. If we have to dump massive amounts of what we know from an enormous variety of different sources to reason about how the world would be different if a hypothesis were false, then that makes the hypotheses an unlikely candidate as the source of any problem the future might bring our way, and very difficult to drop from our reasoning. In an unreflective way we might remark, "Well, that just *couldn't* be false!"

The constructivist suggestion concerning modality is based in what we know. It is what is usually called *epistemic* possibility. Hypotheses are possibly false, given what we know, when we can use that knowledge to conceive of their falsehood. That is, we can still keep quite a lot of our knowledge if we dispense with the outcomes that confirm them, and take the hypothesis to be false. Hypotheses are necessary to the degree that, given what we know of the world, we cannot conceive of their falsehood. That is, it's very difficult to see what would still be so if they are in fact false, even imagining that we are wrong about quite a lot.

To differentiate this approach from others, I will refer to hypotheses as *factually* possible, necessary, and impossible.

We use modal reasoning in designing experiments and reasoning about potential errors, but our reasoning guides choices, and so requires empirical justifications for hypotheses about the world. If something is impossible, we do not need to investigate whether it happened. If something is necessary, we need do nothing to bring it about, and can do nothing to prevent it. If some error is possible then that suggests a control for an experiment, to check that it isn't in fact so. If something might not have become so for some reason, we can sometimes use that information to prevent that kind of thing from becoming so next time, or to bring it about. In this way we can use modal theses, in the practice of science, to reason about whether or not to act and how to bring results about.

Apart from using counterfactuals to diagnose errors, constructivism has some capacity to account for other kinds of counterfactuals which we more commonly meet in philosophy, where we are not concerned with the truth or falsehood of the antecedent. We can often "add" and "subtract" outcomes to confirm counterfactuals. One set of data can, for example, confirm that humans hunt megafauna wherever they live. Another set dates human arrival in Australia, and yet another set documents the vanishing of its megafauna shortly afterwards. Given all this as background, we can confirm that if humans had not arrived in Australia, the megafauna would still be there. Keep the hypothesis that humans hunt megafauna, and subtract out the evidence justifying human arrival in Australia. The remaining evidence makes it probable the megafauna still exist. Australia had a great variety of megafauna that survived ice-ages and other major events for millions of years. So, even aside from the coincidence in dates, no rival account of the extinction is available.

Factual possibility fits some intuitions about modality, but not all. The easiest hypotheses to reason about counterfactually are particular physical events with no widespread consequences for what we observe. It is easy to drop the outcomes that justify them, and the outcomes we have observed that are consequences of their occurrence, because we can retain so much else. Laws of nature are difficult to drop. But some things do not fit very well – the big bang is necessary on this view. One can also object to factual possibility by arguing that general relativity must be factually necessary even though we are able to reason about what would be so if Newton had been correct about gravity. The argument for factual possibility is not that it fits our intuitions well, but that it is a rough and ready model of the way we actually put counterfactual reasoning to use in guiding choices of actions and reasoning about errors.

Factual possibility differs from the notion we are used to in philosophy. It restricts the range of things that are thought of as possible. We can imagine things to be true that are not factually possible. An insignificant example is the Harry Potter stories, which are vividly imagined but in deep conflict with nearly all we know about the world. A more serious one is the allegation that it is always possible to confirm any hypothesis, because we can always *imagine* something that confirms it. Well yes. If we imagine discovering that the speed of light varies with direction, then *perhaps* we could get evidence for Newton's absolute space (Friedman 1983, 6, 115; van Fraassen 1980, 49). I do not see that this evidence is even (factually) possible, let alone that our ability to imagine these results should affect our confidence in anything.

All empirical justifications are defeasible. Even the factually necessary ones are vulnerable to outcomes of observation. So, being casual and dropping the jargon word 'factual', we find that some things that are necessary could be false. That sounds deeply confused. But properly re-phrased, it is a familiar point. Our views about what is necessary and impossible have changed historically. We are concerned here with evidence and justification. When something is factually necessary, we just cannot see a way that it could be false. Trying to envision the way the world is if it is false not only conflicts with nearly everything we think we know, it also conflicts even if we are wrong about many things. But the fact that we cannot see a way for it to be false is perfectly consistent with our being unaware of evidence and reasoning that shows that it is false. We might run into that evidence in the future, as we have for similar cases in the past. We can know that, too. So, rather than saying that any necessary truth might be false, we can say instead that we are confident that a particular hypothesis is factually necessary, but that certainty is subject to future evidence.

Are metaphysical theses necessary? An empirical hypothesis is necessary to the degree that we cannot reason from its falsehood without losing the rest of our empirical knowledge in addition. Clearly, by this criterion, metaphysical theses are virtual paradigms of necessity. The metaphysical theses we ought to keep are factually necessary, but not overwhelmingly confirmed. We might be unable to avoid them, not because of the way the evidence is, but because they are far-reaching consequences of our overwhelmingly confirmed hypotheses, or of our abilities to engage in confirming activity, or both. Dropping them is profoundly disruptive to our knowledge and the way we are forced to justify it. A great deal of mathematics, then, is factually necessary. We cannot engage in the empirical investigations we do engage in without it.

It could also be that, by finding things out through empirical investigation, we cannot avoid behaving in ways that overwhelmingly confirm that we are certain that something is true. I have already noted the argument that by engaging in empirical investigation in the way scientists do, we act in ways that confirm that we believe in both observable and unobservable objects. The same goes for the hypothesis that I am not a brain in a vat and that induction is usually reliable.

But if there's no evidence justifying metaphysical hypotheses, and they are not justified *a priori*, ought we, then, to say that they are justified at all? As long as the situation is clear, I do not see why it matters whether or not you attach the word 'justified' to metaphysical claims that possess no outcomes of observation that confirm them but in which we believe as a by-product of our empirical investigations. We cannot engage in the practice of finding out about the world without acting in ways that confirm that we believe that some identifiable and very general hypotheses, metaphysical hypotheses, are true. So there's a case for attaching the word 'justified' to these hypotheses.

But, while we can confirm that everyone believes this kind of metaphysical claim, we can be unable to confirm *from observation* that they are true. They are necessary because the current crop of theses we dub 'metaphysical' are consequences of both confirming and refuting many hypotheses we test. But as the example of geometry shows, some hypotheses we once thought were metaphysical turned out to be testable after all, and some of them turned out to be false. So perhaps we should not attach the word 'justified' to the ones we believe now.

I will not defend this proposal concerning modality in any detail. It is a line of thought that is suggested by constructivist reasoning, and makes use of the ability of that reasoning to account for relevance relations in empirical reasoning, and enables us to confirm counterfactuals. I'm aware that the proposal is not worked out in detail, and present it only as something suggested by constructivism.

The literature on possible worlds abundantly illustrates the enormous range of theses about necessity and possibility that many philosophers find intuitively appealing. When we ask what we need to believe in order to engage in the practice of science, or informally know about the world, many of these are wholly idle. When we look at our activities, we need to choose intelligently what to do, and that requires empirical justifications that a thing is possible, or impossible, or necessary. Possible worlds – whatever they are and whether or not they exist – are not very good at illuminating the way we actually use these modalities to reason about the world about us. Constructivism gives a way to understand how we justify modal theses by reasoning from selective subsets of actual outcomes of observations, rearranging the reasoning about observations, and sometimes reasoning from different, imaginary sets of outcomes. There is then a preliminary case for thinking that factual necessity and possibility are all we need to engage in empirical investigation, and that anything beyond it is a rich target for doubts.

Thomas Kuhn motivated doubts like these (2000, 79–83). In a famous paper, Hilary Putnam imagined that there was another planet with a substance that is indistinguishable from water under ordinary circumstances, but which was not H_2O. Kuhn pointed out that this was impossible unless we are wildly mistaken about chemistry and physics (that is, it is factually impossible). We would easily differentiate anything with a different chemical structure from water long before chemistry existed. (Imagine dilute alcohol (ethanol) plays the role of water, for example. Under some circumstances, rainfall would catch fire.)

Perhaps some other example will work better. But that is not the point. The point, rather, is the disturbing possibility that we have internalized the intuitions that Putnam draws upon because it is *factually* necessary that water is H_2O. The reason ordinary people might have the intuition that water just *couldn't* be anything other than H_2O might be because they have some inchoate awareness that this is overwhelmingly confirmed in a way that makes it very difficult to make sense of anything around us if we suppose that it's false. If that is so, then the idea of possible worlds might be a misleading detour.

12.3 A case for intuitionist logic and bottom-up metaphysics

Empiricism has traditionally had a parsimonious metaphysics. On the face of it, constructivism appears to be more generous than most versions of it. The constructivist view is that we can confirm hypotheses about unobservable objects. We can confirm some hypotheses about what would have happened if certain events had not occurred. We can know about a Universe that is indifferent to our values, and to our current beliefs. So it looks superficially as though we do not have to accept many limitations.

This isn't so, however. There is a class of metaphysical hypotheses which it is perfectly possible to do without, and which are quite startling. If we view metaphysics this way, we get a very odd view of the Universe and our place in it, one that appears in the work of Michael Dummett (especially his 1978). Dummett presented his view from the perspective of mathematical reasoning. Constructivism provides us with a new way to understand it, one suggested by empirical reasoning instead.

The constructivist view of metaphysical knowledge begins with observations we have made and the beliefs we need in order to find out about the world about us. It doesn't suggest that we have any metaphysical knowledge deriving from a source peculiar to metaphysics. So this is a kind of bottom-up metaphysics.[21] It begins from our mundane situation, and what we know and do not know empirically. Then it proposes that if we have any metaphysical knowledge, it is at most of those hypotheses that are factually necessary for our knowledge of the empirical world.

---

21 "Bump".

12.3 A case for intuitionist logic and bottom-up metaphysics 205

Given this bottom-up view of metaphysics, we need not accept positions in metaphysics which we can show play no role in our ability to confirm hypotheses about the empirical world. This leaves out a particular chunk of our reasoning, which is associated with classical logic.

Consider the question of whether there are homeothermic (warm-blooded) animals in the Andromeda galaxy. We have no idea about this. We can intelligibly discuss the matter. (There are so many stars, with (likely) so many planets, in that galaxy, that I'd be surprised if there weren't *any* with clear cases of homeothermic animals.) We have good evidence we'd be able to recognize these animals if certain events occurred so that we were presented with them. So there's no good case that we do not understand the meaning of the hypothesis. Still, we do not know whether there are such animals. No outcome of observation we have observed so far has (so far as we know) been the result of the existence of such animals, such that we wouldn't make the observation if they do not exist. Perhaps we will never know. Certainly it looks that way at the moment.

Now consider the hypothesis:

Either there are homeothermic animals in Andromeda or there aren't.[22]

I will call this disjunction 'the cited hypothesis'. We certainly have a very strong intuition that this is true. Is it a consequence of our ability to find out about the world?

Well, certainly it is if we possess a classical logic. The law of the excluded middle is a consequence of that logic, so the hypothesis follows. But is it a necessary condition of our engagement in natural science that we adopt such a logic? Do we behave in a way that requires us to be understood as believing the laws of classical logic? Perhaps there is a case here, but I haven't seen it. Indeed, it seems to me that there's a good case for a negative answer (Mitchell 2003). Perhaps we are required to acknowledge the cited hypothesis because of the mathematics we have to use. It might be that we need classical, and not intuitionist, mathematics to investigate the Universe. Certainly we in fact use classical mathematics in physics. But so far, we haven't seen a case that we cannot manage with intuitionist mathematics only.

But really, this kind of argument is a distracting detour. For the obvious reply is that of *course* there either are, or are not, such animals! Because, if there are, then they are *there*, in Andromeda somewhere! And if there aren't then, because Andromeda obviously exists, it follows that they're not *there*! It's

---

22 If you want to make a fuss about this, pretend I've added 'within our past light-cone'.

this kind of very strong intuition that makes it seem puzzling to worry about classical logic. I suspect that it isn't classical logic that causes the intuition that the cited hypothesis is true, but rather the other way around. Einstein famously remarked that the Copenhagen interpretation of quantum mechanics seemed like something put together by a very intelligent paranoid, and intuitionist logic presents the same impression.

What role, though, does the strong intuition that the animals are either there or not play in our ability to find out about the world about us? I agree that if we are in fact going to check Andromeda for homeothermic animals, then certainly the animals are either there or not. But nobody knows if we will. It's even doubtful whether or not creatures located where we are, with our abilities to affect and be affected by the world, could use those abilities to find out. So the question will recur even here: is it true that either we could find out, or we couldn't?

No difference in *our* behavior or knowledge need result from the truth of the cited hypothesis. Nothing shows, for an arbitrary hypothesis about the physical world, that there need be anything we even (factually) *could* observe that shows that the hypothesis is true. Even if the human race lived forever, there need be no time at which we encounter one or the other of these two different kinds of effects. The idea that we either will or won't visit an identified star in Andromeda, or that we either could or couldn't determine whether the animals live there, is part of the same complex of hypotheses. There is a lot we do not know about the world. One of the things we do not know is whether or not there's always something to know.

The upshot of taking the constructive approach to metaphysics is that we are entitled to adopt an intuitionist logic and not include the cited hypothesis as part of our beliefs. Indeed, if constructivism is true, that seems to be the reasonable policy. I would argue that there's no present evidence that shows that any of us even believe the cited hypothesis. (I will not argue this here. The gist of the argument is that we cannot constructively confirm that anyone has this belief. The success or failure of the tasks we can complete cannot be made to depend upon our believing one way or the other. Any evidence from what we are disposed to say or write involves a cycle of justification (see Mitchell 2003). The non-linguistic tasks we can successfully perform provide a way out of a cycle of mutually confirmed hypotheses about what we believe based on what we say. What we are disposed to say is only evidence for belief if there's a way to test it independently by practical success in some task.)

The cited hypothesis concerns distant objects. Other hypotheses of the same kind concern other matters. Are there events lasting for times that are much shorter than we can presently detect? Or particles with a mean free path of much less than we can currently detect? Was there life prior to the formation of the Milky Way? Will there be life in the Milky Way four billion years from now? We do not know the answers to these questions. One can take the questions a long way. Did you put your left sock on first yesterday, or your right one? Perhaps there, too, our practice of investigation doesn't require that we believe the disjunction. There could be answers to each of these questions, and perhaps there are answers to any question about the empirical world. Perhaps God knows the answers. That does, though, appear to sneak in the view that there is something for God to know (Dummett 1991, 350–351).

There's a very important reaction that philosophers have to this kind of argument that I believe is deeply, and importantly, misplaced.

"According to constructivism, in company with intuitionist mathematics, there's some doubt about whether questions to which we do not know the answer really have answers. But clearly *some* of these questions have answers, as the constructive viewpoint allows. Under that view, when there's a known way to investigate and confirm something, we can find out answers even if we haven't done so yet. So the trouble seems to be that some of the things that we do not know are also things we *couldn't* know, or do not know *in principle*."

"But where is this line between what we can in practice investigate, but haven't yet done so, and what we cannot in principle find out? The fact that we haven't yet actually counted, for example, the number of books in the next room, doesn't prevent us knowing (according to constructivism) that it's either odd or even. What makes the cited hypothesis any different? Constructivism allows that we do know that there's a truth of the matter for hypotheses we know how to find out, since that is what motivates us in our empirical investigations. But what if, without us knowing of it, we are all just about to be obliterated? Then we will never find out certain things that we otherwise could. So, under some circumstances, there's an answer to questions that, as a matter of fact, we cannot find out. Well why doesn't that extend to questions we have no idea how to find out too?"

If we agree that an empirical hypothesis is true when and only when it agrees with reality, then, according to constructivism, we must add something else. It agrees with reality, so far as we can tell, when and only when it is overwhelmingly justified. We cannot find out when a hypothesis agrees with reality by comparing it to some separate thing – reality itself. An empirical hypothesis is known only when we are able to see how large numbers of the different ways we are affected by the world must all be manifestations of the truth of the hypothesis. That is, we know an empirical hypothesis only if it is overwhelmingly confirmed. The observations conjointly attest to a single hypothesis that we cannot avoid.

Of course, if the future will in fact go certain ways, then we will be able to find out whether there are warm-blooded animals in Andromeda, and the whole worry will be settled. But we do not know that the future will go that way. We do not know whether or not it even could. And if you argue that some hypothesis still might be right or wrong in spite of our never being able to differentiate between these two states, the reply is that of course it might. To claim that it *is* either right or wrong, is to insist – I see no alternative – that it must either agree with reality, as reality is in itself, or disagree with it. And we simply do not know that stronger claim. We do not find out whether empirical hypotheses are right or wrong in that way at all.

So it is misplaced to ask where the line is between those empirical hypotheses that are true or false even though we haven't investigated them, and those others that we couldn't find out even if we tried. The very demand for such a line is a product of the view that reality must be one way or the other. We must begin where we actually are right now. There are some things we know, and some things we don't. The line between what we can in practice discover if we set out to, and what we cannot or couldn't discover, is one of those things we do not know. So we are entitled to wonder whether there's a line to be drawn here. Saying that there must be a line between the in-principle-discoverable and the not-in-principle-discoverable is an instance of the move from saying that we do not know something to saying that nonetheless there's something to know. But I just do not see that there must be something to know, although I agree that there might be.

That is why it's misplaced to insist that constructivism be explicit about what the difference is between what we know how to find out and what is in principle inaccessable to us. There are some things you do not know but could find out – for example, the number of books in the next room. But constructivism cannot tell you where the limits are on that kind of inquiry because we do not know what we could find out if circumstances are expanded. What we could find out if our information is augmented in various ways is one of the things we just do not know.

So constructivism does leave us with a parsimonious view of the Universe after all. As I have put it, there's a lot we do not know, and one of these things we are ignorant about is whether there must always be something to know. Because we know some things, because the light does genuinely shine, there's no light for the places where it fades away. Nothing we know now shows us the extent of things beyond the illumination. 'Universe' means 'one truth', but nothing shows us that it can be thought of as a whole.

That is the sense in which the metaphysics suggested by constructivism is bottom-up. It begins with what we can actually do, and the truths we have actually discovered, and the way we do actually discover them. Shoes and ships and sealing-wax are thoroughly safe. So are bacteria, bauxite, Betelgeuse and Bose-Einstein condensates. But I know nothing of the nearest possible world where kangaroos lack tails, nothing of headaches without neurons, and nothing, even, of the physical Universe as a whole. I wonder, then, not merely how these things are, but whether they are at all.

We gather round our little light, the sun, and walk upon the earth that made us. And is that not enough? We are not gods but chimpanzees. Emulate Voltaire and cultivate the garden. Your spouse, your child, and all the things you love are here. And so are the myriad humble things that make up life. As Berkeley said, the horse is in the stable, the books are in the study, as before. So are the molecules in the air, the novae and the works of Shakespeare. It's stunning that we know so much, and all we have done to get here. Constructivism, if widely adopted, would alter the way we see metaphysics, the Universe, and perhaps ourselves. It might not give us all we want, but as Lawrence Sklar once observed, citing both Kant and the Rolling Stones, it gives us all we need (Sklar 1988, 55).

Bibliography

Achinstein, P. (1983) *The Concept of Evidence* Oxford, U.K.: Oxford University Press.
Achinstein, P (2001) *The Book of Evidence* Oxford, U.K.: Oxford University Press.
Austin, J. L. (1962) *Sense and Sensibilia* Reconstructed by G. J. Warnock from Austin's notes. Oxford, U.K.: Oxford University Press.
Blackburn, S. (2002) "Realism: Deconstructing the debate" *Ratio* (New Series) Vol. 15, No. 2 (June), pp. 111–133.
Brown, J. R. (1991) *The Laboratory of the Mind: Thought Experiments in the Natural Sciences* London, U.K.: Routledge.
Brown, H. I. (1993) "A Theory-Laden Observation Can Test the Theory" *The British Journal for the Philosophy of Science*, Vol. 44, No. 3 (September), pp. 555–559.
Braithwaite, R. B. (1953) *Scientific Explanation* Cambridge, U.K.: Cambridge University Press.
Callender, C. (2011) "Philosophy of Science and Metaphysics" in S. French and J. Saatsi (eds.), *Continuum Companion to the Philosophy of Science* New York, New York: Continuum, pp. 33–54.
Callender, C. (2017) *What Makes Time Special* Oxford, U.K.: Oxford University Press.
Carnap, R. (1935 [1996]) *Philosophy and Logical Syntax* Bristol, U.K.: Thoemmes Press.
Carnap, R. (1953) "Testability and Meaning" in H. Feigel and M. Brodbeck (eds.), *Readings in the Philosophy of Science* New York, New York: Appleton-Century-Crofts, pp. 47–92. Originally in *Philosophy of Science* Vol. 3, No. 4 (October), pp. 419–471 and Vol. 4, No. 1 (January), pp. 1–40.
Carnap, R. (1956) "Empiricism, Semantics and Ontology" In *Meaning and Necessity* Chicago, Illinois: The University of Chicago Press, pp. 205–222.
Carnap, R. (1963) "Replies and Systematic Expositions" in P. Schillp (ed.), *The Philosophy of Rudolf Carnap: The Library of Living Philosophers volume 11* La Salle, Illinois: Open Court, pp. 859–999.
Carnap, R., and Gardner, M. (1966) *An Introduction to the Philosophy of Science*. New York, New York: Basic Books.
Carrier, M. (1989) "Circles without Circularity" in J. R. Brown and J. Mittelstrauss (eds.), *An Intimate Relation* Dordrecht, North Holland: Klwer Academic Publishers, pp. 405–428.
Chalmers, A. (1976 [1999]) *What is This Thing Called Science?* 3rd edition Indianapolis, Indiana: Hackett Publishing Company.
Chalmers, A. (2003) "The Theory-Dependence of the Use of Instruments in Science" *Philosophy of Science* Vol. 70, pp. 493–509.
Chalmers, D. (2011) "Revisability and Conceptual Change in "Two Dogmas of Empiricism"" *The Journal of Philosophy* Vol. 58, No. 8 (August), pp. 387–415.
Chang, H. (2001) "How to Take Realism beyond Foot-stamping" *Philosophy* Vol. 76, No. 295 (January), pp. 5–30.
Chang, H. (2001a) "Spirit, Air, and Quicksilver: The Search for the "Real" Scale of Temperature" *Historical Studies in the Physical and Biological Sciences* Vol. 31, No. 2, pp. 249–284.
Chang, H. (2004) *Inventing Temperature* Oxford, U.K.: Oxford University Press.
Christensen, D. (1983) "Glymour on Evidential Relevance" *Philosophy of Science* Vol. 50, pp. 471–481.

Christensen, D. (1990) "The Irrelevance of Bootstrapping" *Philosophy of Science* Vol. 57, pp. 644–662.
Christensen, D. (1997) "What is Relative Confirmation?" *Nous* Vol. 31, pp. 370–384.
Christensen, D. (1999) "Measuring Confirmation" *The Journal of Philosophy* Vol. 46, No. 9 (September), pp. 437–461.
Churchland, P. (1985) "The Ontological Status of Observables: In Praise of the Superempirical Virtues" in Paul Churchland and Clifford Hooker (1985), pp. 35–47.
Churchland, P., and Hooker, C. (1985) *Images of Science* Chicago, Illinois: The University of Chicago Press.
Collins, H. (1985) *Changing Order: Replication and Induction in Scientific Practice* London, U.K.: Sage Publications.
Collins, H. (2004) *Gravity's Shadow* Chicago, Illinois: The University of Chicago Press.
Compton, A. H., and Allison, S. K (1935) *X-rays in Theory and Experiment*, New York, New York: D. Van Nostrand Company, Inc.
Compton, A.H. (1973) *Scientific Papers of Arthur Holly Compton* Chicago, Illinois: The University of Chicago Press.
Copernicus, N. (2002 [1543]) *On the Revolutions of the Heavenly Spheres*. Edited by Steven Hawking. Philadelphia, Pennsylvania: Running Press.
Corliss, W. R. (1979) *Mysterious Universe: A Handbook of Astronomical Anomalies* Glen Arm, Maryland: The Sourcebook Project.
Creath, R. (1990) *Dear Carnap, Dear Van*. Berkeley and Los Angles, California: The University of California Press.
Darwin, C. (1859 [1965]) *On The Origin of Species*. Cambridge, Massachusetts: Harvard University Press.
Dorling, J. (1979) "Bayesianism Personalism, the Methodology of Scientific Research Programmes and Duhem's Problem" *Studies in the History and Philosophy of Science* Vol. 10, pp. 177–187.
Duhem, P. (1914 [1991]) *The Aim and Structure of Physical Theory* Princeton, New Jersey: Princeton University Press.
Dummett, M. (1978) *Truth and Other Enigmas* Cambridge, U.K.: Duckworth Press.
Dummett, M. (1991) *The Logical Basis of Metaphysics* Cambridge, Massachusetts: Harvard University Press.
Earman, J. (ed.) (1983) *Minnesota Studies in the Philosophy of Science*. Volume 10, Testing Scientific Theories. Minneapolis, Minnesota: University of Minnesota Press.
Earman, J. (1989) *World Enough and Space-Time* Cambridge, Massachusetts: MIT Press.
Earman, J. (1992) *Bayes or Bust?* Cambridge, Massachuesetts: MIT Press.
Earman, J., and Glymour, C. (1988) "What Revisions Does Bootstrap Testing Need?" *Philosophy of Science* Vol. 50, pp. 260–264.
Edidin, A. (1981) "Glymour on Confirmation" *Philosophy of Science* Vol. 48, No. 2, pp. 292–307.
Edidin, A. (1988) "Discussion: From Relative Confirmation to Real Confirmation" *Philosophy of Science* Vol. 55, pp. 265–271.
Edwards, A. W. F. (1969) "Statistical Methods in Scientific Inferences" *Nature* Vol. 222, pp. 1233–1237.
Edwards, A. W. F. (1972) *Likelihood* Baltimore, Maryland: The Johns Hopkins University Press.

Einstein, A. (1949) "Autobiographical Notes" in P. A. Schilpp (ed.) *Albert Einstein Philosopher Scientist*. Vol. VII, The Library of Living Philosophers. La Salle, Illinois: Open Court Publishers, pp. 3–98.
Friedman, M. (1974) "Explanation and Scientific Understanding" *The Journal of Philosophy* Vol. 71, No. 1. (January 17), pp. 5–19.
Friedman, M. (1983) *Foundations of Space-Time Theories* Princeton, New Jersey: Princeton University Press.
Friedman, M. (2001) *Dynamics of Reason* Stanford, California: CSLI Publications.
Galison, P. (1987) *How Experiments End* Chicago, Illinois: The University of Chicago Press.
Garwin, R., and Levine, J. (1973) "Absence of Gravity-Wave Signals in a Bar at 1695 Hz" and "Single Gravity-Wave Detector Results Contrasted with Previous Coincidence Detection" *Physical Review Letters* Vol. 31, No. 3, pp. 173–176 and 176–80.
Goldman, A. I. (1978) "Epistemics: The Regulative Theory of Cognition" *The Journal of Philosophy* Vol. 75, No. 10, pp. 509–523.
Gillies, D. (1993) *Philosophy of Science in the Twentieth Century* Cambridge, Massachusetts: Basil Blackwell.
Glymour, C. (1980) *Theory and Evidence* Princeton, New Jersey: Princeton University Press.
Glymour, C. (1980hd) "Hypothetico-Deductivism Is Hopeless" *Philosophy of Science* Vol. 47, No. 2 (June), pp. 322–325.
Glymour, C. (1983) "Revisions of Bootstrap Testing" *Philosophy of Science* Vol. 50, No. 4 (December), pp. 626–629.
Gosse, E. (1890) *The Life of Philip Gosse* London, U.K.: Kagen Paul, Trench, Trübner & Company.
Greenwood, J. D. (1990) "Two Dogmas of Neo-Empiricism: The "Theory-Informity" of Observation and the Quine-Duhem Thesis" *Philosophy of Science*, Vol. 57, No. 4 (December), pp. 553–574
Grice, P. (1989) *Studies in the Way of Words* Cambridge, Massachusetts: Harvard University Press.
Grimes, T. (1987) "The Promiscuity of Bootstrapping" *Philosophical Studies* Vol. 51, pp. 101–107.
Grunbaum, A. (1960) "The Duhemian Argument" *Philosophy of Science* Vol. 27, pp. 75–87.
Haack, S. (1993) *Evidence and Inquiry* Oxford, U.K.: Basil Blackwell.
Haack, S. (1998) *Manifesto of a Passionate Moderate* Chicago, Illinois: The University of Chicago Press.
Hacking, I. (1983) *Representing and Intervening* Cambridge, U.K.: Cambridge University Press.
Hacking, I. (1999) *The Social Construction of What*? Harvard, Massachusetts: Harvard University Press.
Hanson, N. R. (1958) *Patterns of Discovery* Cambridge, U.K.: Cambridge University Press.
Harding, S. G. (1976) *Can Theories Be Refuted?* Dordrecht, Holland: D. Reidel Publishing.
Hempel, C. (1965) *Aspects of Scientific Explanation* New York, New York: The Free Press.
Hempel, C. (1966) *Philosophy of Natural Science* Englewood Cliffs, New Jersey: Prentice-Hall, Inc.
Hetherington, N. (1983) "Just how Objective Is Science?" *Nature* Vol. 306, pp. 727–730.
Horwich, P. (1983) "Explanations of Irrelevance" in Earman (1983), pp. 55–66.
Horwich, P. (1991) "On the Nature and Norms of Theoretical Commitment" *Philosophy of Science* Vol. 58, pp. 1–14.

Howson, C., and Urbach, P. (1989) *Scientific Reasoning: The Bayesian Approach*. La Salle, Illinois: Open Court.
Hudson, R. (1994) "Background Independence and the Causation of Observations" *Studies in the History and Philosophy of Science* Vol. 25, No. 4, pp. 595–612.
Hudson, R. (2014) *Seeing Things* Oxford, U.K.: Oxford University Press.
Jeffrey, R. (1965 [1983]) *The Logic of Decision* Chicago, Illinois: The University of Chicago Press.
Kitcher, P. (1978) "Theories, Theorists, Theoretical Change" *The Philosophical Review* Vol. 87, No. 4 (October), pp. 519–547.
Kitcher, P. (1993) *The Advancement of Science* Oxford, U.K.: Oxford University Press.
Kitcher, P. (2001) *Science Truth and Democracy* Oxford, U.K.: Oxford University Press.
Kornblith, H. (2014) "Is there Room for Armchair Theorizing in Epistemology?" in M. Haug (ed.) *Philosophical Methods: The Armchair or the Laboratory?* Abingdon, U.K.: Routledge, pp. 195–216.
Kosso, P. (1988) "Dimensions of Observability" *British Journal for the Philosophy of Science* Vol. 39, pp. 449–467.
Kosso, P. (1989) "Science and Objectivity" *The Journal of Philosophy* Vol. 86, No. 5, pp. 245–257.
Kosso, P. (1989a) *Observability and Observation in Physical Science* Dordrecht, North Holland: Kluwer Academic Publishers.
Knobe, J., and Nichols, S. (2017) "Experimental Philosophy" in: E. N. Zalta (ed.), *The Stanford Encyclopedia of Philosophy* <https://plato.stanford.edu/archives/win2017/entries/experimental-philosophy/>. Retrieved at 9th May 2020.
Kuhn, T. (1957) *The Copernican Revolution* Cambridge, Massachusetts: Harvard University Press.
Kuhn, T. (1962 [1996]) *The Structure of Scientific Revolutions* Chicago, Illinois: The University of Chicago Press.
Kuhn, T. (1977) *The Essential Tension* Chicago, Illinois: The University of Chicago Press.
Kuhn, T,. (2000) *The Road since Structure* Chicago, Illinois: The University of Chicago Press.
Ladyman, J., and Ross, D. (2007) *Every Thing Must Go* Oxford, U.K.: Oxford University Press.
Lakatos, I. (1978) *The Methodology of Scientific Research Programmes*. Volume 1, Philosophical Papers. Edited by John Worral and Gregory Curry. Cambridge, U.K.: Cambridge University Press.
Lakatos, I., and Musgrave, A. (eds.) (1970) *Criticism and the Growth of Knowledge* Cambridge, U.K.: Cambridge University Press.
Laudan, L. (1965) "Grunbaum on 'the Duhemian Argument'" *Philosophy of Science* Vol. 32, pp. 295–301.
Laudan, L. (1996) *Beyond Positivism and Realism* Boulder, Colorado: Westview Press.
Leplin, J. (1997) *A Novel Defense of Scientific Realism* Oxford, U.K.: Oxford University Press.
Lewis, D. (1973) *Counterfactuals* Cambridge, Massachusetts: Harvard University Press.
Lloyd, E. (1983 [1994]) *The Structure and Confirmation of Evolutionary Theory* Princeton, New Jersey: Princeton University Press.
Longino, H. (2001) *The Fate of Knowledge* Princeton, New Jersey: Princeton University Press.
Longino, H. (1990) *Science as Social Knowledge* Princeton, New Jersey: Princeton University Press.
Mach, E. (1893) *The Science of Mechanics* La Salle, Illinois: Open Court Publishing.
Maddy, P. (2007) *Second Philosophy* Oxford, U.K.: Oxford University Press.

Maher, P. (1993) *Betting on Theories* Cambridge, U.K.: Cambridge University Press.
Maher, P. (2004) "Probability Captures the Logic of Scientific Confirmation" in C. Hitchcock (ed.) *Contemporary Debates in Philosophy of Science* Oxford, U.K.: Blackwell, pp. 69–91.
Maudlin, T. (2007) *The Metaphysics Within Physics* Oxford, U.K.: Oxford University Press.
Maxwell, G. (1962) "The Ontological Status of Theoretical Entities" *Minnesota Studies in the Philosophy of Science* Vol. 3, pp. 3–27.
Mayo, D. G. (1996) *Error and the Growth of Experimental Knowledge*. Chicago, Illinois: The University of Chicago Press.
Mitchell, S. (1988) "Constructive Empiricism and Anti-Realism" *PSA* Vol. 1, pp. 174–180.
Mitchell, S. (1995) "Towards a Defensible Bootstrapping" *Philosophy of Science* Vol. 62, pp. 241–260.
Mitchell, S. (2003) "Bivalence as an Issue in Confirmation Theory" *The Philosophical Forum* Vol. 32, No. 2 (Summer), pp. 189–222.
Mitchell, S. (2016) "A Reply to Nina Emery" *Philosophy of Science* Vol. 86, No. 4 (October), pp. 794–806.
Nagel, J. (2000) "The Empiricist Conception of Experience" *Philosophy* Vol. 75 (July), pp. 345–376.
Nagel, E. (1979) *The Structure of Science* Indianapolis, Illinois: Hackett Publishing.
Newcomb, S. (1882) "Discussion of Observed Transits of Mercury" *Astronomical Papers* Washington, Maryland: Bureau of Navigation, Navy Department.
Newcomb, S. (1906) *Side-Lights on Astronomy* New York, New York: Harper & Brothers.
Newton, I. (1689 [1934]) *Mathematical Principles of Natural Philosophy* Translated by A. Motte Revised by F. Cajori. Berkeley, California: The University of California Press.
Pickel, B., and Schulz, M. (2018) "Quinean Updates: In Defense of 'Two Dogmas'" *The Journal of Philosophy* Vol. 65, No. 2 (February), pp. 57–91.
Poincaré, H. (1902 [1952]) *Science and Hypothesis* New York, New York: Dover Publications.
Popper, K. (1959) *The Logic of Scientific Discovery* London, U.K.: Routledge.
Popper, K. (1963) *Conjectures and Refutations* New York, New York: Basic Books.
Psillos, S. (1999) *Scientific Realism: How Science Tracks the Truth* Oxford, U.K.: Routledge.
Ptolemy [=Claudius Ptolemeus] (1984 [141]) *Almagest*. Translated by G.J. Toomer. New York, New York: Springer-Verlag.
Quine, W. V. O. (1969) *Ontological Relativity and Other Essays* New York, New York: Columbia University Press.
Quine, W. V. O. (1953 [1980]) "Two Dogmas of Empiricism" in *From a Logical Point of View: 9 Logico- Philosophical Essays* 2nd edition. Cambridge, Massachusetts: Harvard University Press, pp. 20–46.
Quinton, A. (1962) "Spaces and Times" *Philosophy* Vol. 37, pp. 130–147.
Reichenbach, H. (1958) *The Philosophy of Space and Time* New York, New York: Dover Publications.
Royall, R. (1997) *Statistical Evidence a Likelihood Paradigm* New York, New York: Chapman & Hall/CRC.
Scheffler, I. (1982) *Science and Subjectivity* 2nd edition Indianapolis, Indiana: Hackett Publishing.
Scow, B. (2011) "Does Temperature Have a Metric Structure?" *Philosophy of Science*, Vol. 78, No. 3 (July), pp. 472–489.
Sellars, W. (1963 [1991]) *Science, Perception and Reality* Atascadero, California: Ridgeview Publishing Company.

Shapere, D. (1982) "The Concept of Observation in Science and Philosophy" *Philosophy of Science* Vol. 49, pp. 485–525.
Shiba, K. (1932) "The Most Probable Values of e, e/ m,and h" *Scientific Papers of Physical and Chemical Research* Vol. 19, pp. 97–121.
Simon, H. (1970) "The Axiomatization of Physical Theories" *Philosophy of Science* Vol. 37, No. 1 (March), pp. 16–26.
Sklar, L. (1985) *Philosophy and Spacetime Physics* Berkeley, California: The University of California Press.
Sklar, L. (1988) "Ultimate Explanations: Comments on Tipler" *PSA* Vol. 2, pp. 49–55.
Sober, E. (1999) "Testability" *Proceedings and Addresses of the American Philosophical Association* Vol. 73, pp. 47–76.
Stevens, S. (1946) "On the Theory of Scales of Measurement" *Science* Vol. 103, pp. 677–680.
Strevens, M. (2012) *"Notes on Bayesian Confirmation Theory"* http://www.strevens.org/bct/. Retrieved September 2017
Suppe, F. (1977) *The Structure of Scientific Theories* 2nd edition Urbana and Chicago, Illinois: The University of Illinois Press.
Van Fraassen, B. (1980) *The Scientific Image* Oxford, U.K.: The Clarendon Press.
Van Fraassen, B. (1983) "Glymour on Evidence and Explanation" in J. Earman (ed.) *Minnesota Studies in the Philosophy of Science*. Volume 10, Testing Scientific Theories. Minneapolis, Minnesota: University of Minnesota Press, pp. 165–176.
Van Fraassen, B. (1985) "Empiricism in the Philosophy of Science" in P. Churchland and C. Hooker (eds.) *Images of Science*. Chicago, Illinois: The University of Chicago Press, pp. 245–305.
Van Fraassen, B. (1989) *Laws and Symmetry* Oxford, U.K.: The Clarendon Press.
Van Fraassen, B. (1995) "Against Naturalized Epistemology" in P. Leonardi and M. Santambrogio (eds.) *On Quine: New Essays* Cambridge, U.K.: Cambridge University Press, pp. 68–88.
Van Fraassen, B. (2002) *The Empirical Stance* New Haven, Connecticut: Yale University Press.
van Fraassen, B. (2007) "From a View of Science to a New Empiricism" in B. Monton (ed.) *Images of Empiricism* Oxford, U.K: Oxford University Press, pp. 337–381.
Van Fraassen, B. (2008) *Scientific Representation* Oxford, U.K.: The Clarendon Press.
Weber, J. (1960) "Detection and Generation of Gravity Waves" *Physical Review* Vol. 117, No. 1, pp. 306–313.
Weber, J. (1969) "Evidence for Discovery of Gravitational Radiation" *Physical Review Letters* Vol. 22, No. 24, pp. 1320–1324.
Weber, J., Lee, M., Gretz, D. J., and Rydbeck, R. (1973) "New Gravitational Radiation Experiments" *Physical Review Letters* Vol. 31, No. 12, pp. 779–783.
Weber, J. (1974) "Weber Replies" *Physics Today* Vol. 27, No. 12, pp. 12–13.
Weber, J. (1975) "Weber Responds" *Physics Today* Vol. 28, No. 11, pp. 13 &99.
Worrall, J. (2010) "Theory Confirmation and Novel Evidence" In D. Mayo and A. Spanos (eds.) *Error and Inference* Cambridge, U.K.: Cambridge University Press, pp. 125–154.
Zytkow, J. (1986) "Discussion: What Revisions Does Bootstrap Testing Need?" *Philosophy of Science* Vol. 53, pp. 101–109.

Name Index

Achinstein, Peter 41, 53, 54, 181
Allison, S.K 97, 99, 100, 101, 102, 104
Austin, J.L. 151

Bellarmine, Robert 153, 184
Blackburn, Simon 171
Boerhaave, Herman 115
Bragg, William L. 98–99
Braithwaite, R.B. 3
Brown, J. R. 171

Callender, Craig 193, 195, 196
Carnap, R. 5, 59, 67, 70, 193, 195
Chalmers, Alan 148, 163, 164, 165–167
Chalmers, David 1
Chang, Hasok 19, 76, 111–123, 164
Christensen, David 6, 7, 8, 9, 41, 42, 45, 46, 54
Churchland, Paul 170
Collins, Harold 3, 112, 116
Columbus, Christopher 149
Compton, A.H 97–104, 109, 199
Copernicus, Nikolaus 125, 127, 129–132, 133, 137–143, 157
Corliss, William R. 92
Creath, Richard 5

Darwin, Charles 4, 7, 9, 16, 21–38, 88, 138, 142
Dorling, John 12, 155
Duhem, Pierre 1–6, 8, 9, 11, 15, 21, 31, 39, 41, 52, 59, 63, 83, 88, 89, 90, 93, 109, 110, 125, 126, 139, 145, 146, 147, 152, 155, 156, 161, 167, 179, 180, 181, 183, 191, 192, 194
Dummett, Michael 204, 207

Earman, John 41, 56, 153, 154, 170
Edidin, Aaron 6, 8, 41, 44, 45, 146
Edwards, A.W.F. 14, 26
Einstein, Albert 4, 192, 200, 206, 208

Fahrenheit, Daniel 56
Friedman, Michael 3, 56, 192, 202

Galison, Peter 162
Gardner, Martin 59
Garwin, Richard 116
Gillies, Donald 146
Glymour, Clark 1, 5, 12, 14, 19, 26, 29, 32, 41, 42, 49, 53, 54, 60, 61–64, 86, 125, 126, 127, 133, 141, 191
Gosse, Edmund 142
Greenwood, John D. 165
Grice, Paul 35
Grünbaum, A. 3, 147, 148

Haack, Susan 181, 199
Hacking, Ian 3, 162, 165, 181
Hanson, N.R. 145
Harding, Sandra G. 156, 183, 184
Hempel, Carl 3, 5, 26, 59, 63, 181
Hetherington, Norris 148
Hooker, Clifford 170
Horwich, Paul 8, 171
Hudson, Robert 3, 165
Huxley, Thomas Henry 192

Jeffrey, Richard 14, 153, 154

Kant, Immanuel 192, 193, 209
Kingsley, Charles 142, 184
Kitcher, Philip 3, 150, 157, 181
Kornblith, Hilary 194
Knobe, Joshua 194
Kosso, Peter 75, 165
Kuhn, Thomas 3, 89, 132, 143, 148, 153, 156, 180, 184, 200, 203

Ladyman, James 193, 194
Lakatos, Imre 89, 90, 91, 181
Laudan, Larry 5, 181, 182
Laue, Max von 97, 98

Name Index

Leplin, Jarrett 3, 182
Lescarbault, Edmund 149, 150
Lewis, David 193
Lloyd, Elizabeth 32
Longino, Helen 3, 183, 184

Mach, Ernst 21, 36, 37, 73
Maddy, Penelope 194
Maher, Patrick 26
Maudlin, Tim 194
Maxwell, Grover 75, 145
Mayo, Deborah G. 12, 14, 26, 105, 162
Millikan, Robert 103, 104
Mitchell, Sam 2, 8, 41, 171, 189, 196, 205, 206
Musgrave, Alan 89

Nagel, Ernst 59
Nagel, Jennifer 172, 173
Newcomb, Simon 92
Newton, Isaac 3, 40, 50, 55–58, 89, 90, 91, 92, 122, 143, 192, 194, 201, 202
Nichols, Shaun 194

Pickel, Bryan 1
Poincaré, Henri 69
Popper, Karl 3, 26, 170, 188, 192
Psillos, Stathis 64, 173
Ptolemy, Claudius Ptolemeus 19, 125, 126, 132, 135, 136, 138, 141, 149, 153, 154, 156, 157
Putnam, Hilary 203, 204

Quine, W.V.O. 1–6, 8, 9, 10, 11, 15, 21, 29, 31, 39, 41, 52, 59, 60, 63, 70, 83, 88, 89, 90, 93, 108, 109, 110, 121, 125, 126, 139, 145, 146, 147, 152, 155, 156, 161, 167, 168, 179, 180, 181, 183, 191, 192, 193, 194, 195, 196
Quinton, Anthony 174, 175

Regnault, Henri 115, 121
Reichenbach, Hans 59, 69, 70
Ross, Don 193, 194
Royall, Richard 12, 14

Scheffler, Israel 147, 149, 156
Schlick, Moritz 192
Schulz, Moritz 1
Schwartz, Robert 104
Scow, Bradford 117, 118
Sellars, Wilfrid 53, 145
Shapere, Dudley 75
Shiba, Kamekichi 104
Simon, Herbert 62
Sklar, Lawrence 56, 209
Sober, Elliot 165
Stevens, S. 117
Strevens, Michael 14, 26, 152, 154
Suppe, Fredrick 3

Urbach, P. 12, 26, 50

Van Fraassen, Bas 2, 20, 52, 56, 64, 68–69, 75, 92, 152, 169–177, 180, 181, 185, 186, 189, 194, 202

Weber, Joseph 116, 117
Worrall, John 3

Zytkow, Jan 41, 62

Subject Index

Absolute velocity 55–58
absolute space 40, 55, 56, 57, 58, 202
aether 54, 56
AIDS 42, 43, 44
a *posteriori* knowledge 193, 195
a *priori* knowledgeable 193
analytic-synthetic distinction 1, 59, 179
analytic hypotheses 17, 59
Andromeda 205, 206, 207
Archaeopteryx 44
arctic 21, 22, 27
Australia 201
auxiliary hypotheses 2, 3, 5, 13, 23, 26, 31, 33, 146
Avogadro's number 9, 98, 101, 102, 103, 199

background knowledge 10, 15, 16, 19, 42, 51, 62, 77, 81, 89, 97, 107, 130, 139, 154, 163, 164, 166, 167
backing hypotheses for observation 145, 146, 147, 149, 150, 151, 152, 155, 156, 161, 166, 187
balance 19, 60, 73, 74, 75, 76, 77, 78, 79, 80, 81
bathwater, epistemological 194
battery 62
Bayesian 12, 13, 14, 17, 18, 26, 39, 41n4, 42, 47–51, 60, 62, 65, 66, 67, 70, 152, 153, 154, 155, 166, 167, 170
Bayes' theorem 14, 47, 48, 49, 65, 134, 135
bet, bets 151, 172
bilateral reduction sentences 67, 68
bird 2, 7, 43, 44, 122, 197
– reason fly south 52n6
boiling (water) 113, 114, 115, 117, 119, 120, 121
boiling oil 115, 119
bootstrap theory of confirmation 14, 26, 29, 41, 41n4, 42, 53
bottom-up metaphysics 204–209
boys 31
– birthday-cards 64, 65, 66, 67
– unreliability as wolf-detectors 31
Bragg diffraction 98–99

calculus 111
carbon dioxide 52
cards, birthday 64, 65, 66, 113
Centigrade scale 117
chain of justification 2, 14, 16, 27, 37, 39, 86
Chang's paradox 111–123
Chemistry 73, 102, 203
chimpanzees 9, 188, 209
chi-squared test 104, 105, 107
climate 22, 23, 27, 28, 29, 32, 33, 34, 188
confirmation 1, 5, 6–7, 14, 26, 29, 31, 39, 41n4, 42, 58, 60, 70, 71, 73, 79, 85, 87, 91, 92, 96, 104, 108, 113, 152, 175, 177, 182
– always defeasible 70
– bottom up/ top down 18, 205
– cycles of 15, 18, 55, 56, 113, 125–143, 167
– independent of a hypothesis 40–41
– independent of an observation 40–41
– not incorrigible see *always defeasible* 17
– overwhelming 8, 187, 195, 196, 202, 204, 207
– real 6, 7, 8, 9, 10, 12, 13, 15, 18, 29, 30, 38, 40, 41, 47, 48, 53, 58, 68, 70, 117, 125, 167
– relative 5, 6, 7, 8, 10, 12, 13, 14, 15, 18, 26, 29, 30, 35, 47, 60, 62, 70, 108
– theories of relative 12, 13, 14, 17, 26, 29, 41
conservation of energy 57
conservation of linear momentum 62
consonance 86, 87, 90, 92, 107, 115, 199
constructive Bayesian evidence independence 48
constructive Bayesian hypothesis independence 48
constructive empiricism 2, 69, 172
constructive trees 10, 11, 15, 16, 17, 18, 19, 23n3, 39, 40, 45, 47, 49–51, 70, 72, 81, 82, 83, 84, 86, 106, 107, 125, 129, 130, 131, 133, 146, 152–154, 161, 185, 186, 199
– arcs/edges 15
– in Darwin 21–38

Subject Index

- finite 10, 15, 33
- leaf node 15, 18, 23
- target hypothesis 10, 15, 17, 23, 25, 29, 30, 31, 32, 33, 36, 39, 40, 47, 51, 146, 163, 186
- target node 23n3
- recursive 39
- root node 23n3, 25, 30, 47
- skeleton 23, 81, 82, 133
- "sneak up" as justification 51

constructivism 2, 4, 5, 6, 8, 9, 10, 11, 12, 18, 19, 20, 21, 22, 23, 30, 31, 33, 36, 37, 39, 40, 45, 46, 49, 51, 52, 53, 54, 55, 56, 57, 58, 63, 68, 69–70, 71, 72, 73, 86, 88, 89, 90, 91, 92, 93, 110, 111, 115, 125, 126, 129, 134, 135–143, 145, 147, 152, 153, 154, 155, 156, 159, 161–163, 165, 167, 169, 174, 179, 181, 183, 184, 185–189, 191–209
- defined 2
- and intuitionism 2
- sufficient to practice science 18, 169
- top-down, bottom-up approaches 18, 208

contrast class 52, 53
conservation of energy 57
conservation of linear momentum 62
coordinative definition 59
Copernicanism 126–129, 135–143, 153
copper 118, 119, 120
correspondence rule 59
counterfactuals 140, 198, 201, 203
crystal 98, 99, 100, 101, 102, 103, 109
- ionic spacing in 99, 101
cultural posits 121
cycles of confirmation/justification 15, 18, 19, 39, 53, 54, 55, 56, 63, 72, 91, 113, 125–143, 161, 167, 177, 206
- in observations 161
cycles of anomaly 126, 127, 131–132, 138, 139, 140

Darwin, Charles 4, 7, 9, 16, 21–38, 88, 138, 142
Darwin's finches 7
dephlogisticated air 150
diffraction grating 97, 98, 99, 100, 103

distance *See* length
dog 188
- flea infestations of 188, 189
Doppler effect 164
dreaming 172

earth 4, 9, 11, 27, 88, 116, 126, 127, 128, 129–130, 131–132, 133, 134, 135, 136, 137, 140, 149, 157, 169, 184, 209
ecliptic 127, 130
electro-motive force 62
empiricism 2, 69, 70, 143, 156n18, 169, 170, 171, 172, 183, 191–198, 204
empiricist premise 170, 171
empirically equivalent theories 126, 171
energy 4, 57, 88, 116n12, 117, 118, 119, 120, 169, 172
errors *See* fault-tracing
established background problems 6
evidence independence 48, 50, 164, 166, 167
evidence *See* observation
evolution 9, 31, 96, 142
exchangable bodies on a balance 77, 78
excluded middle, law of 205
experimenter's regress 112, 116
explanation, scientific 52
explanatory connection/integration/ power 181, 182, 188, 193

fault tracing 5, 6, 11, 19, 60, 75, 83–88, 90, 93, 95, 97–104, 107–110, 114, 117, 122, 146, 149, 150, 156, 173, 174, 175, 176, 179, 187, 198, 199
- as a normal part of ordinary science 89
- in Darwin 88
- in length measurement 83
- in polarized sunglasses 93–97
- in temperature measurement 122
- in X-ray diffraction 97–98
- not always successful 173
- orbit of Mercury 89
fixed stars 3, 55, 126, 127, 129, 130, 131, 132, 135, 137–139, 140
flowers, time of blooming 69

Subject Index — 221

forces 45, 52, 54, 55, 62, 73, 81, 92, 161, 182, 188
foundationalism 2, 12, 29, 53, 71, 154, 158, 179
frame of reference 55, 56, 57, 88
freezing (water) 113, 114, 115, 120
fungi 52

gas constant 61
gas laws 61
God, attention of 54
gods, we're not 209
girl, birthday-cards 66, 67
glaciers 22, 27, 28
gloomy thought of future refutation 70
Glymour, bootstrapping 14, 26, 29, 41, 41n4, 42, 53
gravitational field 81, 103
Great Muddle, The 41
Guam 197

Halmasauruses 42, 43, 44
hallucination 20, 169, 172, 173, 174, 176
harmony/harmoniousness 19, 126, 141, 142, 143
Heisenberg uncertainty relations 57
Higgs boson 36, 186
history of science 2, 5, 14, 183
HIV 42, 43
holism 71–72, 146, 191
Hooke's law 62
Hudson, observation 3, 165
human body, as measuring instrument 64, 152, 172
human interests / values 184, 192
hypothesis / hypotheses 1–6, 7, 8, 9, 10, 11, 12, 13, 15, 16, 17, 18, 19, 21, 23, 25, 26, 27, 29, 30, 31, 32, 33, 34, 35, 36, 37, 39, 40, 41, 42, 44–46, 47, 48, 50, 51, 52, 53, 54, 56, 57, 58, 59, 60, 62, 63, 64, 65, 66, 67, 69, 70, 71, 72, 76, 77, 78, 79, 80, 81, 83, 84, 85, 86, 87, 88, 89, 90, 93, 95, 96, 97, 105, 107, 108, 109, 110, 115, 116, 120, 123, 125, 128, 129, 130, 131, 131n15, 133, 134, 135, 139, 141, 142, 145, 146, 147, 148, 151, 152, 153, 154, 155, 156, 157, 158, 161, 162, 163, 164, 165, 166, 167, 172, 174, 175, 176, 179, 180, 181, 183, 184, 185, 186, 187, 188, 189, 191, 192, 193, 194, 195, 196, 197, 198, 199, 200, 201, 202, 205, 206, 207
– analytic 17, 59
– auxiliary 2, 3, 5, 6n1, 13, 23, 26, 30, 31, 32, 33, 42, 46, 62, 90, 125, 142, 146, 154, 167
– beautiful, slaying of 192
– exculpatory 91
– hypothesis independence 39, 40–41, 44–47, 48, 50, 51, 53, 163, 166
– target 2, 5, 7, 9, 10, 11, 13, 15, 17, 18, 23, 25, 26, 27, 29, 30, 31, 32, 33, 35, 36, 39, 40, 42, 45, 46, 47, 48, 50, 51, 59, 62, 90, 125, 142, 146, 154, 155, 161, 162, 163, 166, 167, 186
hypothesis of observation 152, 153, 157
Hypothetico-deductivism 5, 12, 14, 26, 45

ice ages 27, 188
identity of length 67
identity of weight 76, 77, 84, 119
illusions 20, 169, 172, 173, 174, 175, 176, 187, 192
independent confirmation 6–29, 30, 31, 39, 58, 91, 92, 175
– of length 80–81, 83
– meaning of 39–58
– not symmetric 51
– why possible 8
induction 16, 29, 37, 39, 60, 63, 67, 123, 196, 202
intuitionism / intuitionist logic 2
iron 119, 148

Jupiter 126, 127, 131, 138, 139
justifying virtue 13, 125
justification *See* confirmation

Kelvin's objection to Darwin 4
Kepler's laws 45, 46, 54, 143

language, philosophy of 17, 35, 179
law of the lever 60, 73, 82
length

- eyeball-length 63, 64
- fault tracing in 83
- measurement by ruler 50, 63, 64, 72, 83
- measurement from the balance 83
Likelihood theories of confirmation 13, 26
linear scale 116, 117–123
logical positivism 17
- sneering at 192

Macbeth 174
Mach's *The Science of Mechanics* 21, 36, 73
malaria 69, 169
manatees 149
Mars 7, 46, 126, 127, 128, 131, 138, 139, 140
mass 9, 11, 50, 54, 55, 56, 61, 73, 101, 102, 103, 119, 119n13, 122, 188
Maximizing independent agreement, minimizing conflict, with evidence 17
meaning 17, 39–58, 59, 60, 68, 109, 115, 205
measurement *See* scale
Mercury 3, 50, 89–92, 113, 114–117, 118, 120, 121, 122, 127, 131, 135, 136, 137, 149, 150
- metal 119
- planet 91, 127
metaphysics 179, 191–209
metric scale 117, 118
modal reasoning 198, 200, 203
mouse 152, 197
Michelson-Morley 4
molecular weight 98, 101, 102, 103
Monte-Carlo simulation 106
moon 3, 137, 140, 149
mosquito bites 69
Mount Holyoke College 104, 105
mountains 21, 22, 26, 71
Müller-Lyer illusion 174
murder 73, 96

necessity 53, 198, 202, 203
neo-Newtonian spacetime 56
Newton 3, 40, 50, 55–58, 89, 90, 91, 92, 122, 143, 192, 194, 201, 202
Newtonian mechanics 3, 50, 56, 89, 90, 91, 122
Neyman-Pearson theories of confirmation 12

nomic measurement, problem of 111
novelty 64, 87, 181

observable entities 2, 18, 59, 173
- correlations among 64, 69, 72
observation
- candlelight, by 153
- illusory 176
- novel 64
- observation-independence 40–44, 48, 81, 91, 163
- outcomes of 1, 2, 5, 10, 11, 12, 14, 17, 18, 19, 21, 23, 36, 37, 39, 40, 52, 56, 59, 61, 62, 63, 64, 67, 68, 70, 71, 83, 88, 107, 108, 112, 121, 122, 123, 125, 142, 143, 145, 146, 163, 166, 168, 169, 171, 176, 180, 183, 184, 185, 186, 187, 188, 189, 191, 192, 195, 199, 202, 203
- theory-laden 2, 5, 12, 19, 37, 59, 145–159
Ohm's law 62
one leg, standing on 52
ontological principles 121
operationalism 63
order of planets 135, 137, 138–139, 140
ordinal and metric scales 117, 118, 119, 120
original acquisition problems 60, 69, 112, 114
overwhelming justification/confirmation 8, 9, 10, 200, 204, 207
oxpeckers 44
oxygen 52, 150

parsimony 60
phlogiston 54, 149, 150, 158, 159, 185
physicists, more reliable than salespeople 95
pig, detection of 151
planets 9, 45, 46, 54, 89, 91, 126, 127, 131–132, 134, 135, 136, 137, 138–139, 140, 142, 148, 149, 203, 205
Polaris 55, 126, 127
polarized light 94
police 97
possibility 50, 57, 60, 72, 93, 130, 131, 183, 192, 198–204
possible worlds 193, 198–204, 208

Subject Index — 223

practice of science 1, 16, 36, 73, 158, 183, 201, 203
pragmatic virtues 179–182, 185–189, 193, 194
pragmatism in science 2
pregnancy tests 186
Ptolemaic theory 19

quantum mechanics 189, 196, 206
Quine, W.V.O. 1–6, 8, 9, 10, 11, 15, 21, 29, 31, 39, 41, 52, 59, 60, 63, 70, 83, 88, 89, 90, 93, 108, 109, 110, 121, 125, 126, 139, 145, 146, 147, 152, 155, 156, 156n18, 161, 167, 168, 179, 180, 181, 183, 191, 192, 193, 194, 195, 196
– Quine's first dogma 3, 59
– Quine's second dogma 1, 2, 191, 193
– "Two Dogmas of Empiricism" 1, 191, 193
Quine-Duhem Hypotheisis 1
– defined 1

random 106, 107, 199
raven, ravenfeather 41, 42, 43, 44
real confirmation/justification 6, 7, 8, 9, 10, 12, 13, 15, 18, 29, 30, 38, 40, 41, 46, 48, 53, 58, 68, 70, 125, 167
recursive 39
reduction, reductionism 67, 68
– reply to 108
red-shift 51
refutation 1, 11, 12, 13, 26, 40, 49, 53, 72, 86, 89, 90, 137, 146, 156, 171, 182, 186, 196
relative confirmation 6, 7, 8, 10, 12, 13, 14, 17, 18, 26, 29, 35, 41, 47, 60, 62, 70, 108
relative space 55
relative velocity 56, 57
relativity, theory of 88
– general 50, 91
– special 4, 164, 201
relevance problem 12
restoring consonance 86, 87, 90, 92, 107, 115, 199
revolution of longitude 126, 127, 128, 131, 132, 139, 140

rhinos 44
Rolling Stones, The 209

Salem witchcraft trials 105
Santa, pants of 52n6
Saturn 126, 127, 131, 138, 139, 140
scales of justice See balance
scale 78–79, 84, 118, 122, 161
– of length 68, 80–81, 83, 84
– of temperature 118
– of weight 83, 84, 87
science-worship 192
Science 1, 2, 3, 5, 8, 9, 11, 13, 14, 15, 16, 17, 18, 20, 21, 26, 31, 35, 36, 37, 38, 39, 40, 51, 53, 57, 59, 60, 68, 70, 72, 73, 83, 86, 88, 89, 92, 93, 96, 107–110, 111, 121, 122, 141, 143, 145, 146, 147, 155, 156, 156n18, 158, 161, 168, 169, 171, 172, 173, 177, 179, 180, 181, 182, 183, 184, 185, 186, 187, 189, 191, 192, 193, 194, 195, 196, 197, 199n20, 201, 203, 205
– case for verisimilitude 95, 195
– education 14, 35, 36
– practice of 1, 5, 16, 18, 31, 35, 36, 38, 73, 83, 93, 108, 121, 125, 146, 155, 157, 158, 162, 169, 171, 180, 181, 182, 183, 201, 203
– textbooks 16
– worship 192
"secret weight" 85
sense-data 2, 30, 154, 173
shellfish 32
shopping carts 69
single value, principle of 121, 122
Sirius 55, 148
skeptic about unobervables 63, 65
skepticism 53, 65, 66, 125
'smustifications' 72
social and political values 183, 184
solar year 126, 127, 129–130, 134, 135, 140
solar-system
– center of mass 9, 11, 55, 56, 127
– order of planets 132, 137, 138–139, 140
species 4, 21, 22, 26, 27, 28, 29, 30, 31, 32, 34, 44, 122, 145, 200
speed of light 4, 116, 202

sunglasses 93–97, 199
statistical mechanics 117
symmetries, of theory 73, 76

telescope 9, 19, 89, 126, 143, 150, 153, 154, 155, 156, 161
temperature 4, 61, 70, 117, 118, 119, 120, 121, 122, 164
– field 70
– measurement of 112, 113–117, 121
theory-ladenness of observation 5, 145–159
thermometry 19
truth, true 1, 2, 3, 4, 7, 8, 9, 12, 13, 15, 17, 19, 21, 26, 29, 30, 45, 46, 53, 54, 55, 59, 60, 62, 63, 66, 71, 75, 77, 78, 79, 81, 88, 91, 95, 105, 109, 122, 123, 125, 126, 133, 134, 135, 141, 142, 145, 147, 148, 149, 150, 152, 153, 154, 156, 163, 166, 169, 170, 172, 174, 176, 181, 183, 184, 186, 187, 188, 191, 192, 194, 195, 196, 197, 198, 199, 200, 201, 202, 203, 205, 206, 207, 208

ultraviolet catastrophe 4, 88
underdetermination 11
unification 181, 185, 186, 187
Universe 129, 140, 169, 174, 191, 192, 195, 196, 197, 198, 204, 205, 208, 209

– expansion of 51
unobservables 2, 19, 20, 59–63, 64, 66, 67, 68, 69, 71, 73, 75, 76, 81, 112, 113, 118, 137, 169, 170, 171, 172, 176, 196, 202, 204
Uranus 9, 148

values in science 193
van Fraassen's paradox 169–177
Venus 46, 127, 131, 135, 136, 137, 153, 154, 156
verificationism, reply to 108–110
vicious circularity 163–165, 176, 195, 224
voltage 62, 103

wavelength 97–104, 109
weight 9, 19, 73, 74, 75, 76, 77–79, 80, 81, 84, 85, 86, 87, 98, 101, 102, 103, 118, 119, 120, 122
– identity of 77–78, 84
– scale of 83, 84, 87
whales 200

X-ray diffraction 2, 98, 100, 102

young women, delight in men being symbolically hanged 106

www.ingramcontent.com/pod-product-compliance
Lightning Source LLC
Chambersburg PA
CBHW030649230426
43665CB00011B/1015